Torts Law

6th Edition

DAVID GREEN
BARRISTER

Cavendish
Publishing
Limited

London • Sydney • Portland, Oregon

Sixth edition first published in Great Britain 2005 by
Cavendish Publishing Limited, The Glass House,
Wharton Street, London WC1X 9PX, United Kingdom
Telephone: + 44 (0)20 7278 8000 Facsimile: + 44 (0)20 7278 8080
Email: info@cavendishpublishing.com
Website: www.cavendishpublishing.com

Published in the United States by Cavendish Publishing
c/o International Specialized Book Services,
5804 NE Hassalo Street, Portland,
Oregon 97213-3644, USA

Published in Australia by Cavendish Publishing (Australia) Pty Ltd
45 Beach Street, Coogee NSW 2034, Australia

First edition 1993
Second edition 1995
Third edition 1998
Fourth edition 2001
Fifth edition 2003
Sixth edition 2005

British Library Cataloguing in Publication Data
Green, David, 1935 –
Torts – 6th ed – (Q&A series)
1 Torts – England – Examinations, questions, etc
2 Torts – Wales – Examinations, questions, etc
I Title
346.4'2'03

Library of Congress Cataloguing in Publication Data
Data available

ISBN 1-85941-961-5
ISBN 978-1-859-41961-8

1 3 5 7 9 10 8 6 4 2

Printed and bound in Great Britain

Preface

The many changes that have occurred in the law of tort since the earlier editions of this work were published have necessitated a thorough revision throughout the book. Much new material has been added to topics such as occupiers' liability, the application of the Human Rights Act 1998 to nuisance, *Rylands v Fletcher* (1868), breach of statutory duty and causation. However, the author has kept in mind that, as stated in the introduction, the ideal reaction of the reader should be 'I can do that – I know those cases', rather than to marvel at the erudition of the author.

The reception given to the earlier editions of this book suggests that students are in sympathy with its aims and find that it is of use in their battle with the examiners. The author is happy to be judged by these criteria.

The author has attempted to state the law as of January 2005.

David Green
London
January 2005

Torts Law

6th Edition

Contents

Preface v
Table of Cases ix
Table of Legislation xxiii
Introduction xxv

1 Vicarious Liability 1

2 Negligence – Duty of Care Generally and Restricted Situations 13

3 Negligence – Breach, Causation and Remoteness of Damage 43

4 Breach of Statutory Duty 53

5 Employers' Liability 63

6 Product Liability 75

7 Occupiers' Liability 87

8 Nuisance 101

9 The Rule in *Rylands v Fletcher* and Fire 121

10 Animals 137

11 Defamation 147

12 Trespass to the Person, to Land and to Goods 163

13 Economic Torts 177

14 Remedies 189

15 General Defences 197

Index 207

Table of Cases

A v National Blood Authority [2001] 3 All ER 289 . . . 77, 85
AMF International v Magnet Bowling [1968] 1 WLR 1028 . . . 90, 92, 97
Abouzaid v Mothercare (UK) Ltd [2001] 1 All ER (D) 2436 . . . 77, 85
Acrow v Rex Chainbelt [1971] 1 WLR 1676 . . . 179, 182
Adams v Ursell [1913] 1 Ch 269 . . . 106, 110, 116
Addie v Dumbreck [1929] AC 358 . . . 93, 97, 168
Akenzua v Secretary of State for the Home Department [2003] 1 All ER 35 . . . 26
Al Saudi Banque v Clarke Pixley [1989] 3 All ER 361 . . . 21, 22
Alcock v Chief Constable of South Yorkshire Police [1991] 4 All ER 907 . . . 28–36
Alcock v Wraith [1991] EGCS 137 . . . 10, 68
Alderson v Booth [1969] 2 QB 216 . . . 168
Alexandrou v Oxford [1993] 4 All ER 328 . . . 39, 41
Allen v Flood [1898] AC 1 . . . 185
Allen v Gulf Oil Refining [1981] AC 1004 . . . 123, 202
Ancell v McDermot [1993] 4 All ER 355 . . . 39
Andreae v Selfridge and Co [1938] Ch 1 . . . 104, 117, 133
Andrews v King (1991) The Times, 5 July . . . 10
Angel v Bushell and Co [1968] 1 QB 813 . . . 148
Angus v Clifford [1891] 2 Ch 449 . . . 182
Anns v Merton London Borough Council [1978] AC 728 . . . 13–16, 59
Archer v Brown [1985] QB 401 . . . 193
Armory v Delamirie (1721) 1 Stra 505 . . . 170
Ashton v Turner [1981] QB 137 . . . 199, 200, 204
Associated British Ports v TGWU [1989] 1 WLR 939 . . . 179, 182
Associated Newspapers Group v Insert Media Ltd [1988] 1 WLR 509 . . . 178, 184
Aswan Engineering Establishment v Lupdine [1987] 1 All ER 135 . . . 78, 82
Atkinson v Newcastle Waterworks (1877) 2 Ex D 441 . . . 54
Attia v British Gas [1988] QB 304 . . . 33
Attorney General v Cole [1901] 2 Ch 205 . . . 103, 133
Attorney General v Cooke [1933] Ch 89 . . . 139, 142, 145
Attorney General v PYA Quarries [1957] 2 QB 169 . . . 107, 111, 123, 131
Atwood v Small (1838) 6 Cl & Fin 232 . . . 182

B v Islington Health Authority [1991] 1 All ER 825, HC;
(1992) 142 NLJ 565, CA . . . 2, 5, 9, 44, 47, 56, 64, 67, 83, 198
Baggs v UK (1987) Commission Report, 8 July . . . 104, 113, 118
Bamford v Turnley (1860) 3 Bes 62 . . . 103, 107, 110, 116
Barber v Somerset County Council [2004] 2 All ER 385 . . . 69, 71, 72
Barkway v South Wales Transport Co Ltd [1950] AC 185 . . . 2
Barnes v Nayer (1986) The Times, 19 December . . . 165
Barnett v Chelsea and Kensington Hospital Management Committee [1969] 1 QB 428 . . . 48, 50

Barrett v Enfield London Borough Council [1999] 3 WLR 79, HL; [1997] 3 WLR 628, CA. 60, 61
Bell v Department of Health and Social Security (1989) The Times, 13 June 95
Bellew v Cement Co [1948] IR 61 . . . . 106, 110, 116
Belmont Finance Corp v Williams Furniture [1980] 1 All ER 393. . . . . 178, 181
Berkoff v Burchill [1996] 4 All ER 1008 159
Bisset v Wilkinson [1927] AC 177 182
Blyth v Birmingham Waterworks (1856) 11 Ex 781. . . . . 2, 6, 9, 44, 56, 64, 67, 81, 83, 89, 90, 108, 112, 133, 198
Bognor Regis Urban District Council v Campion [1972] 2 QB 169 158
Bolam v Friern Hospital Management Committee [1957] 1 WLR 582 . . . . 9, 48, 83
Bolitho v City and Hackney Health Authority [1997] 4 All ER 771 49
Bolton v Stone [1951] 1 All ER 1078 . . . . 103, 106, 115, 127, 130, 132, 145
Bonnick v Morris [2002] 3 WLR 820. . . . . 148, 159, 160
Borgion v Ministry of Agriculture [1985] 3 All ER 585 25
Bowater v Rowley Regis Corp [1944] KB 476 . . . . 199, 203
Bradburn v Great Western Railway (1874) LR 10 Ex 1 . . . . 191, 195
Bradford v Robinson Rentals [1967] 1 WLR 337 144
Branson v Bower (2001) The Times, 23 July 153
Brice v Brown [1984] 1 All ER 997. . . . . 29, 33, 65
Bridges v Hawkesworth (1851) 21 LJ QB 75. 171
Bridlington Relay v Yorkshire Electricity Board [1965] Ch 436. 104, 115, 127, 133
Brimelow v Casson [1924] 1 Ch 302 186
British Celanese v Hunt [1969] 2 All ER 1252 . . . . 107, 118, 123
British Railways Board v Herrington [1972] AC 877. 399
Buckle v Holmes [1926] 2 KB 125 139
Burmah Oil v Lord Advocate [1965] AC 75 202
Byrne v Deane [1937] 1 KB 818 . . . . 149, 152, 156
Byrne v Hoare (1965) Qd R 135 171

CBS Songs v Amstrad Consumer Electronics plc [1988] Ch 61 58
Calgarth, The [1927] P 93 . . . . 89, 90, 171
Calverley v Chief Constable of Merseyside Police [1989] AC 1228. 38
Cambridge Water Co v Eastern Counties Leather [1994] 2 WLR 55 . . . . 103, 107, 111, 112, 115, 117, 118, 122, 123, 125–27, 130–32, 134
Caparo Industries plc v Dickman [1990] 2 WLR 358; [1990] 1 All ER 568 . . . . 2, 5, 9, 15, 16, 18, 20–23, 44, 47, 56, 59, 64, 67, 83, 90, 108, 112, 133, 198
Capital and Counties Bank v Henty (1882) 7 App Cas 741 152
Capital and Counties plc v Hampshire County Council [1997] 2 All ER 695 . . . . 41, 42
Carslogie Steamship v Royal Norwegian Government [1952] AC 292 52
Cassell v Broome [1972] AC 1027. 194
Cassidy v Daily Mirror [1929] 2 KB 331 . . . . 155, 156, 159
Castle v St Augustine's Links (1922) 38 TLR 615 . . . . 115, 134
Century Insurance v Northern Ireland Road Transport Board [1942] AC 509 . . . . 7, 9, 68
Chadwick v British Transport Commission [1967] 1 WLR 912 . . . . 28, 30, 32, 200, 203

Chapman v Honig [1963] 2 QB 502 . . . 179, 182
Chaudry v Prabhakar [1989] 1 WLR 29 . . . 17, 18, 22
Chester v Afshar [2002] 4 All ER 587 . . . 52
Christie v Davey [1893] 1 Ch 316 . . . 104, 116
Church of Jesus Christ of the Latter Day Saints (GB) v West Yorkshire Fire and Civil Defence Authority (1997) The Times, 20 March . . . 41
Clough v Bussan [1990] 1 All ER 431 . . . 39
Clunis v Camden and Islington Health Authority [1998] 3 All ER 180 . . . 199, 203
Cole v Turner (1704) 6 Mod 149 . . . 173
Coltman v Bibby Tankers [1988] AC 276 . . . 66
Conway v Wimpey [1951] 2 KB 266 . . . 3, 4, 65, 167
Cookson v Knowles [1979] AC 556 . . . 191
Cope v Sharpe [1912] 1 KB 496 . . . 202
Cork v Kirby MacLean [1952] 2 All ER 402 . . . 2, 6, 9, 44, 47, 50, 56, 64, 67, 81, 84, 89, 90, 112, 133, 198
Cory Lighterage v TGWU [1973] ICR 339 . . . 186
Costello v Chief Constable of Northumbria Police [1999] 1 All ER 550 . . . 38
Cowan v Chief Constable of Avon and Somerset Constabulary [2001] EWCA Civ 775 . . . 40
Crofter Hand Woven Harris Tweed v Veitch [1942] 1 All ER 147 . . . 185
Cross v Kirkby (2000) The Times, 5 April . . . 200, 204
Crossley v Faithfull and Gould Holdings Ltd [2004] 4 All ER 447 . . . 69
Cruickshank Ltd v Chief Constable of Kent County Constabulary (2002) The Times, 27 December . . . 186
Cummings v Grainger [1977] QB 397 . . . 139, 141
Cunard v Antifyre [1933] 1 KB 551 . . . 111, 117
Curran v Northern Ireland Co-Ownership Housing Association [1987] AC 718 . . . 15
Curtis v Betts [1990] 1 All ER 769 . . . 141, 169
Cutler v United Dairies [1933] 2 KB 297 . . . 200
Cutler v Vauxhall Motors [1971] 1 QB 418 . . . 50
Cutler v Wandsworth Stadium [1949] AC 398 . . . 54, 58

D and F Estates v Church Commissioners [1989] AC 177 . . . 10, 68, 78, 82
D v East Berkshire Community Health Trust [2003] 2 All ER 796 . . . 61
DC Thomson v Deakin [1952] Ch 646 . . . 179
Dakhyl v Labouchere [1908] 2 KB 325 . . . 149
Dann v Hamilton [1939] 1 KB 509 . . . 7, 199, 200, 201, 203
Darby v National Trust [2001] EWCA Civ 189 . . . 96
Davidson v Chief Constable of North Wales [1994] 2 All ER 597 . . . 166, 175
Davie v New Merton Board Mills [1959] AC 604 . . . 66
Davies v Swan Motor Co [1949] 2 KB 291 . . . 200
De Beers Products v Electric Co of New York [1975] 1 WLR 972 . . . 179, 183
Derbyshire County Council v Times Newspapers [1993] 2 WLR 449 . . . 158, 161
Derry v Peek (1889) 14 App Cas 337 . . . 182
Deyoung v Stenburn [1946] KB 227 . . . 69

Donoghue v Folkestone Properties Ltd [2003] 3 All ER 1101 . . . . . 90, 93, 95, 97, 98
Donoghue v Stevenson [1932] AC 562 . . . . . 2, 5, 9, 14, 15, 44, 47, 56, 64, 67, 75, 77, 80, 81, 83, 84, 89, 90, 95–97, 133, 198
Dubai Aluminium Co Ltd v Salaam [2003] 1 All ER 97 . . . . . 3, 7
Dulieu v White [1901] 2 KB 669 . . . . . 7, 28, 29, 32, 45, 48, 65, 84, 112, 141, 144
Duncan v British Coal Corp [1997] 1 All ER 540 . . . . . 30, 36

EC Commission v UK [1997] All ER (EC) 481 77
Edgington v Fitzmaurice (1885) 29 Ch D 459 182
Edwards v Lee (1991) 141 NLJ 8 . . . . . 18, 22
Edwards v Railway Executive [1952] AC 737 89
Edwards v West Hertfordshire General Hospital Management Committee [1957] 1 WLR 415 69
Edwin Hill v First National Finance Corp [1989] 1 WLR 225 186
Ellis v Sheffield Gas Consumers Co (1853) 2 E & B 767 . . . . . 10, 68
Emeh v Kensington, Chelsea and Westminster Area Health Authority [1985] 2 WLR 233 14
Emerald Construction v Lowthian [1966] 1 WLR 691 186
Erven Warnink v Townend [1979] AC 731 187
Esso Petroleum v Mardon [1976] 1 QB 801 . . . . . 17, 22, 23
Euro-Diam v Bathurst [1988] 2 All ER 23 . . . . . 94, 199, 204
Evans v Triplex Safety Glass [1936] 1 All ER 283 . . . . . 66, 78, 81

F v West Berkshire Health Authority [1989] 2 All ER 545 173
Fairchild v Glenhaven Funeral Services [2002] EWCA Civ 878 . . . . . 43, 52
Ferguson v Welsh [1987] 3 All ER 777 . . . . . 88, 89, 92, 97
Filliter v Phippard (1847) 11 QB 347 . . . . . 134, 135
Fitter v Veal (1701) 12 Mod 542 . . . . . 189, 195
Froom v Butcher [1975] 3 WLR 379 . . . . . 4, 197, 200
Frost v Chief Constable of South Yorkshire Police [1997] 1 All ER 540 . . . . . 28, 29, 35
Fytche v Wincanton Logistics plc [2004] 4 All ER 221 55

Gaca v Pirelli General plc [2004] 3 All ER 348 195
General Engineering Services v Kingston and St Andrews Corp [1989] 1 WLR 69 . . . . . 4, 7, 65, 72
Giles v Walker (1890) 24 QBD 656 . . . . . 122, 125, 129
Gitsham v Pearce [1992] PIQR P 57 . . . . . 64, 69, 92
Glasgow Corp v Taylor [1922] 1 AC 44 96
Godfrey v Demon Internet Ltd [1998] 4 All ER 342 161
Goldman v Hargrave [1967] 1 AC 645 . . . . . 117, 134
Goldsmith v Bhoyrul [1997] 4 All ER 268 158
Gorris v Scott (1874) LR 9 Exch 125 . . . . . 55, 56, 73
Gould v McAuliffe [1941] 2 All ER 527 90
Gran Gelato v Richcliffe [1992] 1 All ER 865 . . . . . 17, 22
Greater Nottingham Co-op v Cementation Piling and Foundations [1989] QB 71 . . . . . 78, 82
Greatorex v Greatorex [2000] 1 WLR 1970; [2000] 4 All ER 769 . . . . . 30, 36
Green v Chelsea Waterworks (1894) 70 LT 547 123

Green v Fibreglass Ltd [1958] 2 All ER 521 . . . 9
Green v Goddard (1702) 2 Salk 641 . . . 164
Greers v Pearman and Corder (1922) 39 RPC 406 . . . 183
Groves v Lord Wimborne [1898] 2 QB 402 . . . 55, 73
Guerra v Italy (1998) 26 EHRR 357 . . . 108, 118, 124

H v Norfolk County Council [1997] 1 FLR 384 . . . 60
Hadmor Productions v Hamilton [1982] 2 WLR 322 . . . 179, 182
Halsey v Esso Petroleum [1961] 2 All ER 145 . . . 102, 107, 110, 130
Hannah v Peel [1945] KB 509 . . . 171
Harris v Birkenhead Corp [1976] 1 WLR 279 . . . 90
Harris v Empress Motors [1983] 3 All ER 56 . . . 190
Harrison v British Railways Board [1981] 3 All ER 679 . . . 200
Harrison v Rutland [1893] 1 QB 142 . . . 168
Harrison v Southwark and Vauxhall Water Co [1891] 2 Ch 409 . . . 102–04, 110, 115, 132
Hartt v Newspaper Publishing (1989) The Times, 9 November . . . 148, 159
Haseldine v Daw [1941] 2 KB 343 . . . 11, 68, 81, 89, 93, 97
Hatton v Sutherland [2002] 2 All ER 19 . . . 69, 72
Hatton v UK [2003] All ER (D) 122 (July) . . . 104, 108, 109, 113, 118, 119
Haynes v Harwood [1935] 1 KB 146 . . . 200
Heasmans v Clarity Cleaning Co Ltd [1987] IRLR 262 . . . 4, 6, 7, 65, 71
Heath v Mayor of Brighton (1908) 98 LT 718 . . . 111
Heaven v Pender (1883) 11 QBD 503 . . . 14
Hedley Byrne v Heller [1964] AC 465 . . . 17, 18, 20, 22, 23
Heil v Rankin [2000] 3 All ER 138 . . . 191
Hellwig v Mitchell [1910] 1 KB 609 . . . 150, 168, 174
Henderson v Jenkins and Sons [1970] AC 282 . . . 123, 131
Henderson v Merrett Syndicates [1994] 3 WLR 761 . . . 18, 23
Herald of Free Enterprise, Re The (1989) The Guardian, 2 May . . . 29, 33
Hill v Chief Constable of West Yorkshire [1989] AC 53 . . . 15, 26, 39
Holden v Chief Constable of Lancashire [1987] QB 380 . . . 194
Hollywood Silver Fox Farm v Emmett [1936] 2 KB 468 . . . 104, 116, 127
Holtby v Brigham and Cowan (Hull) Ltd [2000] 3 All ER 421 . . . 50, 52
Home Office v Dorset Yacht [1970] AC 1004 . . . 14, 59
Honeywill and Stein Ltd v Larkin Bros [1934] 1 KB 191 . . . 10, 68
Horrocks v Lowe [1975] AC 135 . . . 153, 159, 160
Hotson v East Berkshire Area Health Authority [1987] AC 50 . . . 52
Howard Marine v Ogden [1976] QB 574 . . . 17, 22, 23
Hudson v Ridge Manufacturing [1957] 2 QB 348 . . . 71, 72
Hughes v Lord Advocate [1963] AC 837 . . . 6, 9, 56, 65, 67, 199
Hughes v National Union of Mineworkers [1991] 4 All ER 278 . . . 38, 42
Hughes v Waltham Forest Health Authority (1990) The Times, 9 November . . . 49
Hulton v Jones [1910] AC 20 . . . 156
Hunter v British Coal Corp [1998] 2 All ER 97 . . . 30, 36

Hunter v Canary Wharf Ltd [1997] 2 All ER 426, HL; [1996] 1 All ER 482, CA;
(1994) The Independent, 20 December, HC 102, 104, 106, 107, 111, 112, 115, 118, 123, 125, 127, 131, 133, 145
Hussain v New Taplow Paper Mills [1988] AC 514. 195

ICI v Shatwell [1965] AC 656 . . . . 3, 199, 203
Interlink Express Parcels Ltd v Night Trunkers Ltd [2001] EWCA Civ 360. 10
Irving v The Post Office [1987] IRLR 289. . . . . 4, 6, 7, 65, 71, 72

Jaggard v Sawyer [1995] 2 All ER 189. 103
Jefford v Gee [1970] 2 QB 130 191
Joel v Morrison (1834) 6 C & P 501. 6
John Munroe (Acrylics) v London Fire and Civil Defence Authority [1997] QB 983. 41
John v Mirror Group Newspapers Ltd [1996] 2 All ER 35 196
Johnson v BJW Property Developments Ltd [2002] 3 All ER 574 . . . . 134, 135
Jolley v Sutton London Borough Council [2000] 3 All ER 409, HL;
[1998] 3 All ER 559, CA. 96
Jones v Boyce (1816) 1 Stark 493. . . . . 56, 144, 145, 200
Jones v Livox Quarries Ltd [1952] 2 QB 608 . . . . 56, 200
Jones v Swansea City Council [1990] 3 All ER 737. 25
Joyce v Sengupta [1993] 1 WLR 337. 161
Junior Books v Veitchi [1983] AC 520 . . . . 13, 14, 25, 78, 82, 123

Kay v Ayrshire and Arran Health Board [1987] 2 All ER 417 52
Kearns v General Council of the Bar [2003] 2 All ER 534 . . . . 153, 165, 168, 175
Kelley v Dawes (1990) The Times, 27 September 190
Kemsley v Foot [1952] AC 345. . . . . 153, 157
Kennaway v Thompson [1981] QB 88 . . . . 111, 113, 117
Kent v Griffiths [2000] 2 All ER 474 42
Khodaparast v Shad [2000] 1 All ER 545. 162
Khorasandjian v Bush [1993] 3 WLR 477 . . . . 102, 106, 111, 115, 127, 133, 145
Kirkham v Chief Constable of Greater Manchester Police
[1990] 3 All ER 246 . . . . 16, 19, 40, 203
Kite v Napp (1982) The Times, 1 June. 141
Knight v Home Office [1990] 3 All ER 237. . . . . 49, 108, 112
Knightley v Johns [1982] 1 WLR 349. . . . . 38, 45, 48, 51, 85, 142, 145
Knowles v Liverpool City Council [1993] 4 All ER 321. 66
Knupffer v London Express Newspapers [1944] AC 116. . . . . 149, 156
Kralj v McGrath [1986] 1 All ER 54 193
Kubach v Hollands [1937] 3 All ER 907. 84
Kuddus v Chief Constable of Leicestershire Constabulary
[2001] UKHL 29; [2001] 2 WLR 1789; [2001] 3 All ER 193 194

Lane v Holloway [1968] 1 QB 379 . . . . 150, 174
Latham v Johnson and Nephew Ltd [1913] 1 KB 395 . . . . 88, 96

Latimer v AEC [1953] AC 643 . . . . . 64, 68, 69, 72
Law Society v KPMG Peat Marwick [2000] 1 WLR 1921; [2000] 4 All ER 540 22
Leach v Chief Constable of Gloucestershire Constabulary [1998] 1 All ER 215 40
League Against Cruel Sports v Scott [1985] 2 All ER 459 167
Leakey v National Trust [1980] QB 485 117
Leigh and Sillivan v Aliakmon Shipping [1985] 1 QB 350, CA; [1988] AC 785, HL 15
Lennon v Metropolitan Police Commissioner [2004] 2 All ER 266 . . . . . 18, 23
Letang v Cooper [1965] 1 QB 232 173
Lewis v Daily Telegraph [1964] AC 234 148
Liddle v Yorkshire (North Riding) County Council [1934] 2 KB 101 . . . . . 88, 96
Lim Poh Choo v Camden and Islington Area Health Authority [1980] AC 174 . . . . . 190, 191
Limpus v London General Omnibus (1862) 1 H & C 526 . . . . . 3, 65
Linklater v Daily Telegraph (1964) 108 SJ 992 156
Lippiatt v South Gloucestershire Council [1999] 4 All ER 149 . . . . . 111, 113
Lister v Hesley Hall Ltd [2001] UKHL 22; [2001] 2 WLR 1311 . . . . . 2, 3, 7
Lloyd v Grace, Smith and Co [1912] AC 716 7
London Artists v Littler [1969] 2 QB 375 . . . . . 149, 152, 157, 160
Lonrho v Al-Fayed [1991] 3 WLR 188 . . . . . 179, 182, 185, 186
Lonrho v Shell Petroleum [1982] AC 173 . . . . . 54, 58, 179, 182, 185
López Ostra v Spain (1995) 20 EHRR 277 . . . . . 104, 113, 118, 124
Loutchansky v Times Newspapers Ltd [2001] All ER (D) 207 160
Lumley v Gye (1853) 2 El & Bl 216 . . . . . 179, 185
Lyne v Nichols (1906) 23 TLR 86 183
Lyon v Daily Telegraph [1943] KB 746 150

McCamley v Cammell Laird Shipbuilders [1990] 1 All ER 854 195
McCullagh v Lane Fox and Partners (1994) The Times, 25 January 182
McDermid v Nash Dredging and Reclamation [1987] AC 906 . . . . . 55, 64, 68, 69, 71, 72
McFarlane v EE Caledonia Ltd [1994] 2 All ER 1 . . . . . 29, 33, 36
McGhee v National Coal Board [1973] 1 WLR 1 52
McKenna v British Aluminium Ltd (2002) The Times, 25 April 108, 123, 127, 130
McKew v Holland and Hannen and Cubitts [1969] 3 All ER 1621 . . . . . 19, 51, 56, 144, 164
McKinnon Industries v Walker (1951) 95 SJ 559 . . . . . 103, 111, 116
McLoughlin v O'Brian [1983] AC 410 . . . . . 14, 28, 30–32, 34
McManus v Beckham [2002] EWCA Civ 939; [2002] 1 WLR 2982 . . . . . 157, 159
McNaughton (James) Paper Group v Hicks Anderson [1991] 1 All ER 134 21
McQuire v Western Morning News [1903] 2 KB 100 157
Mahon v Osborne [1939] 2 KB 14 45
Malone v Laskey [1907] 2 KB 141 . . . . . 102, 103, 106, 107, 111, 115, 123, 133, 145
Marc Rich and Co AG v Bishop Rock Marine Co Ltd [1995] 3 All ER 307 . . . . . 15, 16
Marcic v Thames Water Utilities Ltd [2001] 3 All ER 698 . . . . . 109, 118, 124
Marcic v Thames Water Utilities Ltd [2004] 1 All ER 135 . . . . . 108, 113, 118
Market Investigations v Minister of Social Security [1969] 2 QB 173 10
Marks v Chief Constable of Greater Manchester (1992) The Times, 28 February 193

Marquess of Bute v Barclays Bank [1955] 1 QB 202 170
Martin v Owen (1992) The Times, 21 May 192, 195
Mason v Levy Auto Parts [1967] 2 QB 530 134
Mason v Williams and Williams [1955] 1 WLR 549 . . . . . 66, 78, 81, 82
Matthews v Ministry of Defence [2003] 1 All ER 689 61
Mattis v Pollock [2004] 4 All ER 85 . . . . . 3, 7
Maynard v West Midlands Health Authority [1985] 1 All ER 635, HL 48, 49
Maynegrain v Campafina Bank [1984] 1 NSWLR 258 . . . . . 168, 170
Merivale v Carson (1887) 20 QBD 275 149
Merkur Island Shipping v Laughton [1983] 2 AC 570 . . . . . 179, 182, 185, 186
Metropolitan Saloon Omnibus v Hawkins (1859) 4 H & N 87 158
Michaels v Taylor Woodrow Developments Ltd [2001] Ch 493; [2001] 2 WLR 224 . . . . . 178, 185
Middlebrook Mushrooms Ltd v Transport and General Workers Union (1993) The Times, 18 January 186
Middleweek v Chief Constable of Merseyside [1990] 3 All ER 662 166
Midwood v Manchester Corp [1905] 2 KB 597 . . . . . 106, 115, 130, 145
Miller v Jackson [1977] 3 WLR 20 . . . . . 112, 113, 117
Milne v Express Newspapers Ltd [2003] 1 All ER 482 . . . . . 156, 161
Mirvahedy v Henley [2004] 3 All ER 401 . . . . . 137–39, 141, 169
Mogul Steamship v McGregor Gow [1892] AC 25 . . . . . 178, 185
Moorgate Mercantile v Twitchings [1977] AC 890 170
Morgan v Fry [1968] 3 All ER 452 186
Morgan v Girls Friendly Society [1936] 1 All ER 404 . . . . . 10, 68
Morgan v Odhams Press [1971] 1 WLR 1239 . . . . . 156, 159
Morris v Murray [1990] 2 WLR 195 . . . . . 7, 199, 203
Mulcahy v R (1868) LR 3 HL 306 . . . . . 178, 185
Murphy v Bradford Metropolitan Council [1992] PIQR P 68 . . . . . 92, 95
Murphy v Brentwood District Council [1990] 2 All ER 908 . . . . . 2, 5, 9, 15, 44, 47, 56, 64, 67, 83, 89, 90, 108, 112, 133, 198
Musgrove v Pandelis [1919] 2 KB 43 134
Mutual Life and Citizens Assurance v Evatt [1971] AC 793 . . . . . 17, 22

Nash v Sheen (1953) The Times, 13 March 164
National Coal Board v England [1954] AC 403 . . . . . 90, 94, 199, 200, 203
Nettleship v Weston [1971] 2 QB 691 . . . . . 2, 5, 6, 44, 198, 199, 203
Ng Chun Pui v Lee Chuen Tat [1988] RTR 298 . . . . . 45, 123, 131
Nicholls v Rushton (1992) The Times, 19 June 29, 33
Nitrigin Eireann Teoranta v Inco Alloys [1992] 1 All ER 854 . . . . . 78, 82
Normans Bay Ltd v Condert Brothers (2004) The Times, 24 March 52
Nor-Video Services Ltd v Ontario Hydro (1978) 84 DLR (3d) 221 104

OLL Ltd v Secretary of State for Transport [1997] 3 All ER 897 42
Ogwo v Taylor [1987] 2 WLR 988 . . . . . 36, 96, 134
Orange v Chief Constable of West Yorkshire Police [2001] 3 WLR 736 40
Oropesa, The [1943] P 32 . . . . . 45, 48, 51, 85, 139, 142, 145
Osman v Ferguson [1993] 4 All ER 344 . . . . . 39, 40

Osman v UK [1998] 5 BHRC 293. . . . . . 40, 60, 61
Owens v Brimmel [1977] 2 WLR 94 200

Padbury v Holliday and Greenwood (1912) 28 TLR 492 . . . . . . 10, 68
Page v Smith [1995] 2 WLR 644 . . . . . . 29, 35
Paris v Stepney Borough Council [1951] AC 367. 71
Parker v British Airways Board [1982] 1 QB 1004 . . . . . . 170, 171
Parkins v Scott (1862) 1 H & C 153 . . . . . . 150, 174
Parmiter v Coupland (1840) 6 M & W 103 . . . . . . 148, 149, 152, 155
Parry v Cleaver [1970] AC 1. . . . . . 191, 195
Pasley v Freeman (1789) 3 TR 51 182
Peabody Donation Fund v Sir Lindsay Parkinson [1985] AC 210 15
Pearson v Lightning (1998) The Independent, 30 April 16
Performance Cars v Abraham [1962] 1 QB 33. 50
Perry v Kendricks Transport [1956] 1 All ER 154 131
Phelps v Hillingdon London Borough Council [2000] 3 WLR 776; [2000] 4 All ER 504 61
Philips v Whiteley [1938] 1 All ER 566. . . . . . 83, 84
Phipps v Rochester Corp [1955] 1 QB 540 88
Pickett v British Rail Engineering [1980] AC 136 190
Pidduck v Eastern Scottish Omnibus [1990] 2 All ER 69. . . . . . 192, 195
Pinn v Rew (1916) 32 TLR 451 . . . . . . 10, 68
Pitcher v Martin [1937] 3 All ER 918 168
Pitts v Hunt [1990] 3 WLR 542. . . . . . 4, 7, 90, 94, 199, 200, 203, 204
Polemis, Re [1921] 3 KB 560. . . . . . 141, 142, 164
Pride of Derby v British Celanese [1953] Ch 149. . . . . . 111, 117

Quinn v Leatham [1901] AC 495 185

R v Bournewood Community and Mental Health NHS Trust [1998] 1 All ER 634 . . . . . 165, 167, 173, 202
R v Deputy Governor of Parkhurst Prison ex p Hague:
Weldon v Home Office [1992] 1 AC 58 165
R v Howson (1966) 55 DLR (2d) 582. 168
R v Meade and Belt (1823) 1 Lew CC 184 175
R v Self [1992] 1 WLR 657. . . . . . 165, 168, 174
R v Wilson [1955] 1 WLR 493 175
Racz v Home Office [1994] 2 WLR 23 3
Rae v Mars UK (1989) The Times, 15 February 89, 93, 96
Rainham Chemical Works v Belvedere Fish Guano Co [1921] 2 AC 465 126
Rantzen v Mirror Group Newspapers [1993] 3 WLR 953. 196
Ratcliff v McConnell [1999] 1 WLR 670 . . . . . . 90, 97
Ratcliffe v Evans [1892] 2 QB 524 . . . . . . 179, 187
Ravenscroft v Rederiaktiebolaget Transatlantic
[1992] 2 All ER 470, CA; [1991] 3 All ER 73, HC. 15, 31, 35
Read v Hudson (1700) 1 Ld Raym 610 152
Read v Lyons [1947] AC 156 . . . . . . 122, 126, 129, 139, 142, 145
Ready Mixed Concrete v Minister of Pensions [1968] 2 QB 497 10

Redgrave v Hurd (1881) 20 Ch D 1 . . . 182
Reeves v Commissioner of Police of the Metropolis [1998] 3 All ER 897 . . . 16, 40
Reid v Rush and Tomkins [1990] 1 WLR 212 . . . 69, 72
Revill v Newberry [1996] 2 WLR 239; [1996] 1 All ER 291 . . . 90, 200
Reynolds v Times Newspapers Ltd [1999] 4 All ER 609 . . . 160, 161
Richardson v LRC Products Ltd [2001] Lloyd's Rep Med 280 . . . 77, 85
Rickards v Lothian [1913] AC 263 . . . 122, 126, 130, 131, 134
Rigby v Chief Constable of Northamptonshire Police [1985] 1 WLR 1242; [1985] 2 All ER 985 . . . 38
Roberts v Ramsbottom [1980] 1 All ER 7 . . . 44
Robinson v Kilvert (1889) 41 Ch D 88 . . . 103, 106, 111, 116, 130
Robinson v Post Office [1974] 1 WLR 1176 . . . 48, 142, 144
Robson v Hallett [1967] 2 QB 393 . . . 164
Rookes v Barnard [1964] AC 1129 . . . 180, 182, 186, 194
Rose v Miles (1815) 4 M & S 101 . . . 123, 131
Rose v Plenty [1976] 1 WLR 141 . . . 3, 65
Roshner v Polsue and Alfieri Ltd [1906] 1 Ch 234 . . . 107, 116
Royal Bank Trust (Trinidad) v Pampellonne [1987] 1 Lloyd's Rep 218 . . . 18, 22
Ryeford Homes v Sevenoaks District Council (1989) 16 Con LR 75 . . . 118, 123
Rylands v Fletcher (1868) LR 3 HL 330 . . . 101, 106–13, 117, 118, 121–28, 130–32, 134, 137, 139, 142, 145

Sadgrove v Hole [1901] 2 KB 1 . . . 152
Salmon v Seafarer Restaurants [1983] 3 All ER 729 . . . 96
Salsbury v Woodland [1969] 3 All ER 863 . . . 10
Saunders v Edwards [1987] 1 WLR 1186 . . . 94, 204
Scala Ballroom v Ratcliffe [1958] 3 All ER 220 . . . 185
Scott v London and St Katherine's Docks (1865) 2 H & C 596 . . . 45
Scott v Shepherd (1773) 2 Wm Bl 892 . . . 139, 173
Sedleigh-Denfield v O'Callaghan [1940] AC 880 . . . 117, 131
Shakoor v Situ [2000] 4 All ER 181 . . . 84
Shelfer v City of London Electric Lighting Co [1895] 1 Ch 287 . . . 103
Silkin v Beaverbrook Newspapers [1958] 2 All ER 516 . . . 149, 153, 157
Sim v Stretch [1936] 2 All ER 1237 . . . 148, 152, 155, 159, 174
Simaan General Contracting v Pilkington Glass [1988] QB 758 . . . 78, 82
Simms v Leigh Rugby Football Club [1969] 2 All ER 923 . . . 98
Slazengers v Gibbs (1916) 33 TLR 35 . . . 156
Slim v Daily Telegraph [1968] 1 All ER 497 . . . 149, 153, 157, 161
Smith v Baker [1891] AC 325 . . . 47, 55, 66, 72
Smith v Eric Bush [1990] 1 AC 829 . . . 17, 18, 20, 22, 23
Smith v Kenrick (1849) 7 CB 515 . . . 122, 125, 129
Smith v Land and House Prop Corp (1884) 28 Ch D 7 . . . 182
Smith v Leech Brain [1962] 2 WLR 148 . . . 2, 6, 9, 45, 48, 56, 65, 68, 84, 112, 199
Smith v Littlewoods Organisation [1987] AC 241 . . . 98, 131, 145
Smithies v National Association of Operative Plasterers [1909] 1 KB 310 . . . 186

Smoker v London Fire and Civil Defence Authority [1991] 2 All ER 449 191
Smolden v Whitworth (1996) The Times, 18 December. 16
Sorrel v Smith [1925] AC 700. 178
South Staffordshire Water Co v Sharman [1896] 2 QB 44. 171
Southwark Borough Council v Williams [1971] Ch 734 202
Sparrow v Fairey Aviation [1964] AC 1019. 73
Spring v Guardian Assurance plc [1994] 3 WLR 354. . . . . 17, 18, 23, 153, 160
Square Grip Reinforcement v MacDonald 1968 SLT 65 179
St Helen's Smelting Co v Tipping (1865) 11 HL Cas 642 . . . . 104, 106, 116, 123, 130, 133
Stanley v Powell [1891] 1 QB 86 173
Stansbie v Troman [1948] 1 All ER 599 . . . . 69, 145
Staples v West Dorset District Council [1995] PIQR 439. 96
Stapley v Gipsum Mines [1953] AC 663. 49
Steedman v British Broadcasting Corp [2001] EWCA Civ 1534 . . . . 150, 154
Stevenson, Jordan and Harrison Ltd v MacDonald [1952] 1 TLR 101. 10
Stokes v Guest, Keen and Nettlefold [1968] 1 WLR 1776 71
Stone v Taffe [1974] 1 WLR 1575. 98
Storey v Ashton (1869) LR 4 WQB 476 . . . . 3, 6
Stovin v Wise (1996) The Times, 26 July 60
Stratford v Lindley [1965] AC 269 185
Stubbings v Webb [1993] 1 All ER 322 204
Sturges v Bridgman (1879) 11 Ch D 852. . . . . 103, 107, 110, 111, 116, 117, 133
Summers, John v Frost [1955] AC 740 54
Sutherland Shire Council v Heyman (1985) 157 CLR 424. 15
Swinney v Chief Constable of Northumbria Police [1996] 3 All ER 449. . . . . 38, 40

TP and KM v UK [2001] 2 FCR 289. 61
Tarry v Ashton (1876) 1 QBD 314 134
Tate and Lyle v Greater London Council [1983] 1 All ER 1159 202
Taylorson v Shieldness Produce Ltd [1994] PIQR 329 35
Telnikoff v Matusevich [1991] 3 WLR 952 . . . . 149, 153, 157
Tetley v Chitty [1986] 1 All ER 663. . . . . 102, 107, 110, 111, 117
Thackwell v Barclays Bank [1986] 3 All ER 676. . . . . 94, 204
Theaker v Richardson [1962] 1 WLR 151 152
Thomas v Bradbury Agnew [1906] 2 KB 627. . . . . 149, 153, 157
Thomas v NUM (South Wales Area) [1985] 2 All ER 1 . . . . 164, 173
Thompson v Commissioner of Police of the Metropolis [1997] 2 All ER 762. 194
Thomson v Cremin [1956] 1 WLR 103. 97
Thomson v Deakin [1952] Ch 646; [1952] 2 All ER 361. 185
Thornton v Kirklees Metropolitan Borough Council [1979] QB 626 55
Three Rivers District Council v Bank of England (No 3) [2000] 3 All ER 1 24–26
Tinsley v Milligan [1993] 3 WLR 126 . . . . 90, 94, 200, 204
Tomlinson v Congleton Borough Council [2002] EWCA Civ 309. . . . . 90, 93, 95, 98
Toronto Power v Paskwan [1915] AC 734 . . . . 55, 66, 72

Transco v Stockport Metropolitan Borough Council [2004] 1 All ER 589 . . . . . 107, 112, 121–23, 125–30, 134, 139, 142, 145
Twine v Beans Express [1946] 1 All ER 202 . . . . . 3, 4

Vacwell Engineering v BDH Chemicals [1971] 1 QB 88 7
Vanderpant v Mayfair Hotel [1930] 1 Ch 138 . . . . . 103, 133
Vellino v Chief Constable of Greater Manchester [2002] 1 WLR 218 . . . . . 38, 94, 200, 204
Victorian Railway Commissioners v Coulthas (1888) 13 App Cas 222 . . . . . 28, 32
Vowles v Evans [2003] EWCA Civ 318 16

W v Essex County Council [2000] 2 All ER 237 . . . . . 30, 34
W v Meah [1986] 1 All ER 935 193
Wagon Mound (No 1), The [1961] AC 388 . . . . . 2, 6, 7, 9, 19, 45, 48, 56, 64, 65, 67, 81, 84, 89, 90, 108, 112, 133, 198
Wagon Mound (No 2), The [1967] 1 Lloyd's Rep 402 . . . . . 103, 107, 111, 115, 132
Walker v Northumberland County Council [1994] 1 All ER 737 . . . . . 69, 72
Walter v Selfe (1851) 20 CJ Ch 433 . . . . . 102, 103, 106, 110, 115, 130, 132
Walters v North Glamorgan NHS Trust (2003) The Times, 13 January; [2002] All ER (D) 87 (Dec) 35
Walters v Smith [1914] 1 KB 595 . . . . . 165, 168, 173, 174
Waters v Commissioner of Police of the Metropolis [2000] IRLR 720 38
Watkins v Secretary of State for the Home Department [2004] 4 All ER 1158 26
Watt v Longsden [1930] 1 KB 130 . . . . . 153, 154, 160, 165, 168, 174
Weir v Wyper (1992) The Times, 4 May 200
Weldon v Home Office [1991] 3 WLR 341 166
Weller v Foot and Mouth Disease Research Institute [1965] 3 All ER 560 123
Wells v Cooper [1958] 2 All ER 527 84
Wells v Wells [1998] 3 All ER 481 189
West v Shepherd [1964] AC 326 191
Westwood v Post Office [1974] AC 1 56
Whatman v Pearson (1868) LR 3 CP 422 . . . . . 3, 6
Wheat v Lacon [1966] AC 522 . . . . . 88, 90, 92
White v Chief Constable of South Yorkshire Police [1999] 1 All ER 1 . . . . . 28–30, 32, 35, 36
White v Holbrook Precision Castings [1985] IRLR 215 72
White v Jones [1995] 2 WLR 187 . . . . . 18, 23
White v Mellin [1895] AC 154 183
White v St Albans City and District Council (1990) unreported 93
Wieland v Cyril Lord Carpets [1969] 3 All ER 1006 . . . . . 19, 51, 56, 144, 165
Wilkinson v Downton [1987] 2 QB 57 163
Williams v Hemphill Ltd 1966 SLT 259 . . . . . 3, 6
Willson v Ministry of Defence [1991] 1 All ER 638 . . . . . 190, 191
Wilsher v Essex Area Health Authority [1988] 1 All ER 871 52
Wilson v Pringle [1987] QB 237 . . . . . 171, 173
Wilsons and Clyde Coal v English [1938] AC 57 . . . . . 10, 71
Winkfield, The [1902] P 42 170

Woodward v Mayor of Hastings [1945] KB 174 . . . . 11, 89, 93, 97
Woollins v British Celanese (1966) unreported 93
Wright v British Railways Board [1983] 2 AC 773 191
Wright v Lodge [1993] 4 All ER 299 51
X v Bedfordshire County Council [1995] 3 WLR 152 . . . . 54, 57–61

Yewens v Noakes (1880) 6 QBD 530 9
Yorkshire Dale Steamship v Minister of War Transport [1942] AC 691 51
Young v Charles Church (Southern) Ltd (1997) 39 BMLR 146 29
Yuen Kun-Yeu v AG of Hong Kong [1988] AC 175 15
Yukong Line Ltd of Korea v Rendsburg Investments Corp of Liberia [1998] 4 All ER 82 185

Z v UK (2001) 10 BHRC 384 61

Table of Legislation

Statutes

Administration of Justice Act 1982—
- s 1(1) 196
- ss 1(1)(b), 4(2) 191

Animals Act 1971 . . . . . 137–41, 167, 168
- s 2(1) 144
- s 2(2) 138, 139, 141, 142, 168, 169
- s 2(2)(a)–(c) 138, 139, 141, 169
- s 2(3) 138
- s 5(1)–(3) 139, 141, 144
- s 6(2) 138, 141, 143, 144
- s 6(2)(a)–(b) 144
- s 6(3) 138, 141, 144, 169
- s 10 . . . . . 139, 141, 144
- s 11 139

Consumer Protection Act 1987 . . . . . 75–80, 83, 85, 86
- Pt 1 . . . . . 75, 76
- ss 1, 2 76
- s 2(1) 76, 80, 85
- s 2(2) 76, 80
- s 2(3) 76
- s 3 . . . . . 76, 77
- s 3(1) 80, 85
- s 3(2) 76, 80, 85
- s 4 . . . . . 76, 77, 80, 85
- s 4(1) 81
- s 4(1)(d) 78, 80
- s 4(1)(e) 76, 77, 79, 85
- s 5 80
- s 5(2)–(4) 81
- s 7 77

Courts and Legal Services Act 1990—
- s 8 196

Damages Act 1996—
- s 2(1) 190

Defamation Act 1952 155
- s 1 155
- s 2 . . . . . 150, 180
- s 3 181
- s 3(1)(b) 179, 183
- s 5 159

Defamation Act 1996 147
- s 1 . . . . . 158, 161
- s 2 . . . . . 156, 161
- s 2(4) 156, 158
- ss 3, 4 . . . . . 156, 161
- ss 5, 8 . . . . . 150, 154
- s 8(2) 150
- s 9 . . . . . 150, 154

Employers' Liability (Defective Equipment) Act 1969—
- s 1 47
- s 1(1) 55, 66, 72

Factories Act 1961—
- s 14 . . . . . 54, 73

Fatal Accidents Act 1976 . . . . . 191, 193, 195
- ss 3(3), 4 . . . . . 192, 195

Fires Prevention (Metropolis) Act 1774 . . . . . 132, 135
- s 86 134

Guard Dogs Act 1975—
- s 1 139
- s 5 58
- s 5(1) 54, 139

Human Rights Act 1998 . . . . . 57, 58, 60, 61, 101, 104–06, 108–10, 113, 114, 118, 122–24
- s 1 . . . . . 104, 108, 113, 118, 123
- s 2 40
- s 6 . . . . . 101, 104, 109
- s 6(1) 61, 104, 108, 109, 113, 118, 124
- s 6(6) 104, 108, 118, 124

Law Reform (Contributory Negligence) Act 1945—
- s 1(1) 200

Law Reform (Miscellaneous Provisions) Act 1934—
- s 1 191

Law Reform (Personal Injuries) Act 1948—
- s 2(4) 191

Libel Act 1843 150

Limitation Act 1980 . . . 204
ss 2, 11 . . . 189, 204
s 32A . . . 150, 154
s 33 . . . 189, 204
Occupiers' Liability Act 1957 . . . 11, 68, 87, 95–98
s 1(1) . . . 95
s 1(3)(b) . . . 93
s 2(1) . . . 88, 92, 95, 98
s 2(2) . . . 92, 95
s 2(3)(a) . . . 88, 96
s 2(3)(b) . . . 96
s 2(4)(a) . . . 88, 89, 93, 96
s 2(4)(b) . . . 11, 68, 89, 92, 93, 96
s 2(6) . . . 141
Occupiers' Liability Act 1984 . . . 87, 90, 93, 94, 97, 98
s 1(2) . . . 90
s 1(3) . . . 93, 97
s 1(4) . . . 93
s 1(5), (8)–(9) . . . 94

Police and Criminal Evidence Act 1984 . . . 164, 165, 167, 168, 173
s 24(4) . . . 165, 167
s 24(5) . . . 165, 167, 173, 174
s 24(6) . . . 165, 168, 174, 175

Reserve and Auxiliary Forces (Protection of Civil Interests) Act 1951—
s 13(2) . . . 194
Road Traffic Act 1988—
s 4 . . . 199
s 149 . . . 203
s 149(3) . . . 4, 7, 199, 201

Slander of Women Act 1891—
s 1 . . . 156
Social Security (Recovery of Benefits) Act 1997 . . . 191, 195
s 3 . . . 195
Supreme Court Act 1981—
s 32A . . . 190
s 35A . . . 191
s 50 . . . 103

Torts (Interference with Goods) Act 1977—
s 3 . . . 171
s 3(2) . . . 168

Unfair Contract Terms Act 1977 . . . 98, 203
s 1(1)(c), (3) . . . 98
s 2 . . . 4
s 2(1)–(2) . . . 98, 203
s 2(3) . . . 4

Statutory Instruments

General Product Safety Regulations 1994 (SI 1994/2328) . . . 85

EU Legislation

Directives

Directive 85/734/EEC (product liability) . . . 76
Art 7(e) . . . 77

Treaties and Conventions

European Convention on the Protection of Human Rights and Fundamental Freedoms . . . 119
Art 1 . . . 118
Art 2 . . . 101, 108, 118, 123, 124
Art 6 . . . 40, 60, 61
Art 8 . . . 61, 101, 104, 106, 108, 109, 113, 118, 119, 123, 124
European Convention on the Protection of Human Rights and Fundamental Freedoms, Protocol 1—
Art 1 . . . 123

Introduction

The law of tort is a fundamental area of English law and, in addition to being a 'core' subject for the legal profession, a clear understanding of its principles is required for other areas as diverse as employment law and company law. It illustrates, moreover, another characteristic of English law, in that it is primarily a common law area, that is, its rules have been developed through the decisions of the courts rather than being laid down by statute. As a result of this, the student of the law of tort is faced with a bewildering array of cases and rules, and often finds difficulty in deciding what information is relevant to a problem.

This book attempts to help students of tort. It is not intended as a substitute for lectures or for reading standard textbooks or law reports or articles; it is rather aimed at students whose problem is not that they feel that the legal input they have received is insufficient, but rather that it is too great and that they have difficulty in ascertaining what material is essential and what is of lesser importance. A careful study of the answers to the questions contained in this book should reveal those essential areas, and the student will see how basic concepts re-appear not only in questions designed to test that topic in depth, but in other questions which, at first sight, appear to be testing unrelated areas.

Another function of this book is to illustrate how to answer questions in the law of tort. These answers are not intended to be perfect solutions, if such a thing exists. Rather, they are intended to illustrate the sort of well-structured answer that would attract high marks using the knowledge that a well-prepared student should possess. All the cases and principles cited should be familiar to the student – the author is attempting to show how, with the knowledge that the student has, he or she can present it in such a way as to gain the best possible grades. In particular, emphasis has been placed on the way in which the fundamental legal principles relevant to a question should be stated; it is a habit of both authors of suggested solutions and students to hunt through law reports or their minds to find a long-forgotten case which is on all fours with the facts of a question and triumphantly present it as the 'right' answer. As any examination question should be designed to test the student's grasp of legal principles and ability to apply those principles to a factual situation, a moment's reflection will show that although such an approach may point to the correct answer as regards (say) liability, it will not attract the best possible grades. Thus, the author has tried, at all times, to cite cases with which the student should be familiar. An ideal reaction by the reader to the suggested solutions in this book should be 'I can do that', not 'Gosh, how clever the author is; I could never write an answer like that'.

The questions used in this book are typical LLB examination questions, both as regards style and complexity.

This book is not intended to replace lectures or standard textbooks. It is, however, intended to fill a need which, during decades of teaching law, the author has found does exist, namely, to enable the student to gain the best possible examination grades from whatever knowledge he or she possesses. The more reading a student does of lecture notes and standard textbooks, the more he or she should benefit from a study of this book.

1 Vicarious Liability

Introduction

Vicarious liability is a topic which is regularly tested by examiners, either as a question in its own right or as part of a question on (say) negligence or employers' liability. Course of employment and express prohibitions are areas that are especially popular with examiners but, in all vicarious liability questions, it is vital to remember that a primary liability between the tortfeasor and the victim must be established before any liability can be transferred.

Checklist

Students must be familiar with the following areas:

(a) definition of employer and employee;

(b) course of employment and the close connection test laid down in *Lister v Hesley Hall Ltd* (2001);

(c) frolics and detours; and

(d) courts' differing attitudes to careless and deliberate acts.

Question 1

Alpha Manufacturing plc is having its premises decorated by Beta Decorators Ltd. Brian is employed by Beta and needs to collect some additional decoration materials from Beta's premises. He meets Alan, who is employed by Alpha as a driver, and asks Alan if he can give him (Brian) a lift to Beta's premises. Alan agrees and gives Brian a lift in one of Alpha's vans which displays the following notice on the dashboard: 'Only employees of Alpha Manufacturing plc are allowed to travel in this vehicle. Alpha Manufacturing plc accepts no liability whatsoever to any other persons who travel in this vehicle.'

Whilst travelling to Beta's premises, the van hits a lamppost because Alan is adjusting the car radio and not looking where he is going. Brian is thrown out of the van and is severely injured. He was not wearing a seat belt and, if he had been, his injuries would have been much reduced.

Advise Brian.

Answer plan

This is a traditional vicarious liability question that raises various other points of contributory negligence: *volenti* and *res ipsa loquitur*.

The following points need to be discussed:

- liability of Alan to Brian;
- vicarious liability of Alpha for Alan's actions;
- effect of prohibition; and
- contributory negligence on Brian's part.

Answer

It must first be decided whether Brian can sue Alan and, if so, whether Alpha is vicariously liable to Brian. Brian must first show that Alan owed him a duty of care. This poses no problem, as it has been held in a number of cases that a driver owes a duty of care to his passengers (*Nettleship v Weston* (1971)). Where a duty of care has been found to exist previously, there is no need to apply the incremental formulation preferred by the House of Lords in *Caparo Industries plc v Dickman* (1990) or *Murphy v Brentwood District Council* (1990). One could also note the statement of Potts J at first instance in *B v Islington Health Authority* (1991), where he stated that in personal injury cases the duty of care remains as it was pre-*Caparo*, namely, the foresight of a reasonable man (*Donoghue v Stevenson* (1932)), a finding that does not appear to have been disturbed on appeal (1992).

Secondly, Brian must show that Alan was in breach of this duty, that is, that a reasonable person would not have acted in this way (*Blyth v Birmingham Waterworks* (1856)), and this is clear from the facts of the problem. *Res ipsa loquitur* is not applicable here, as the reason for the crash is known (*Barkway v South Wales Transport Co Ltd* (1950)). Brian will have to show that the breach caused him injuries, and the 'but for' test in *Cork v Kirby MacLean* (1952) proves the required causal connection. Finally, Brian will have to prove that the harm he has suffered was not too remote, that is, it was reasonably foreseeable (*The Wagon Mound (No 1)* (1961)). This should cause no difficulty to Brian, because all that he will have to show is that some personal injury was reasonably foreseeable; he will not have to show that the extent was foreseeable (*Smith v Leech Brain* (1962)). Brian could therefore successfully sue Alan.

The next question is whether Alpha is vicariously liable for Alan's negligence. We are told that the employer/employee relationship exists, but is Alan acting in the course of his employment at the relevant time? The traditional approach to this question is to ask whether Alan's act is a wrongful and unauthorised mode of doing an authorised act, or whether it is so unconnected with the authorised act as not to be a mode of doing it, but rather an independent act. However, in *Lister v Hesley Hall Ltd* (2001), the House of Lords reviewed the law on course of employment. The House stated that an employer will also be liable for unauthorised acts provided they are so connected with authorised acts that they may rightly be regarded as modes, albeit improper modes, of those acts. The House held that it was necessary to concentrate on the relative closeness of the connection between the nature of the employment and the tort, and to take a broad approach to the nature of employment. Thus, their Lordships found that an employer was vicariously liable for the sexual abuse of boys who were resident at a boarding school owned by the employer. The House reiterated this approach in *Dubai Aluminium Co v Salaam* (2003), and in *Mattis v Pollock* (2004), the Court of Appeal used the test to find that the employer of a nightclub doorman was liable for an attack carried out on a club patron, despite a large element of personal revenge on the part of the doorman.

It is clear that at the time of the act in question, that is, the negligent driving of Alan, the express prohibition will not automatically take the act outside the course of employment. What the restriction can do is to restrict those acts which lie within the course of employment. However, it cannot restrict the mode of doing an act that does lie within the course of employment: see, for example, *Limpus v London General Omnibus* (1862). Thus, we need to decide whether Alan's act of driving in contravention of the prohibition is an unauthorised act, or whether he is merely carrying out an authorised act in an unauthorised manner. We also need to examine the closeness of Alan's act to the nature of his employment, taking a broad view of the nature of Alan's employment.

In *Limpus*, a driver, contrary to an express prohibition, obstructed a bus from a rival company and caused an accident. It was held that the employers were vicariously liable, because the driver's act was merely a wrongful mode of carrying out an authorised act in the driving of a bus. However, in *Twine v Beans Express* (1946), where a hitchhiker was given a lift, it was held that the driver, by giving a lift to an unauthorised person, was acting outside the course of his employment. *Conway v Wimpey* (1951) is a similar decision involving unauthorised passengers. However, in *Rose v Plenty* (1976), the Court of Appeal disapproved of the decision in *Twine*, which was based, *inter alia*, on the ground that the unauthorised passenger was a trespasser and thus was owed no duty of care. While this was true in 1946, it no longer represents the law since the House of Lords' decision in *British Railways Board v Herrington* (1972) and the Occupiers' Liability Act 1984. Another ground for the decision in *Twine* was based on the duty of care that the employer owed to the passenger, an approach which is no longer correct (see *ICI v Shatwell* (1965)). As a result, the Court of Appeal in *Rose* found it possible to depart from *Twine* and took a broad view of course of employment.

It is submitted that a court would follow *Rose*. In *Rose,* Lord Denning distinguished between two groups of cases, namely, those where the prohibited act was done for the employer's business, when it will usually be held to be within the course of employment (for example, *Limpus*), and those done for some other purpose, for example, giving a lift to a hitchhiker, which if prohibited may lie outside the course of employment (for example, *Twine*; *Conway*). On this analysis, Alan would still be within the course of his employment. Furthermore, in *Racz v Home Office* (1994), the House of Lords also took a wide approach to the concept of course of employment. The decision in *Lister* and *Dubai Aluminium* would enable a court to take a broad view of Alan's employment and find that Alpha is vicariously liable for Alan's negligence.

One way in which Alan could have moved out of the course of his employment would be if he had departed from his authorised route and was on a frolic of his own. The whole area of frolics was considered in *Whatman v Pearson* (1868); *Storey v Ashton* (1869); and by the House of Lords in *Williams v Hemphill* (1966). In the latter case it was held that, to constitute a frolic of his own, the journey in question had to be entirely unconnected with the employer's business. Hence, the fact that Alan is giving a lift to Brian in order that he may collect further decorating supplies to use in decorating Alpha's premises means that the journey was not undertaken entirely for Alan's selfish purposes. Therefore, Alan remains within the course of his employment and Alpha will be vicariously liable for his negligence. While this seems likely, it is by no means certain. The courts have, in recent years, distinguished between careless acts and deliberate acts, and have taken a very narrow view of course of employment where deliberate acts are concerned (see *Heasmans v Clarity Cleaning Co Ltd* (1987); *Irving v Post Office* (1987)). Perhaps the most dramatic example of this approach is to be found in *General Engineering Services v Kingston and St Andrews Corp* (1989), where firemen who drove very slowly to a fire were held not to be within their course of employment in so doing, on the grounds that they were employed to travel to the scene of the fire as quickly as reasonably possible. In travelling as slowly as possible, they were not doing an authorised act in an unauthorised manner; rather, they were doing an unauthorised act. In view of the fact that Alan's giving of a lift to Brian was a deliberate act, a court might feel inclined to follow *Conway* or *Twine*.

In addition, Alpha's notice will be subject to s 2 of the Unfair Contract Terms Act (UCTA) 1977, so that, to the extent that it purports to exclude liability for death or personal injury, it is void.

Alan and thus Alpha could invoke the defence of contributory negligence against Brian. Brian's failing to wear a seat belt has contributed to the extent of his injuries and, therefore, his damages would be reduced by 15% (*Froom v Butcher* (1975)).

The defence of *volenti* may also be raised against Brian for accepting a lift despite the notice in the van. For this to succeed, Brian must have submitted voluntarily to the risk of injury. This seems most unlikely, especially as Brian was unaware of any danger when he accepted the lift from Alan. Also, by s 2(3) of the UCTA 1977, the awareness of the notice is not of itself to be taken as indicating voluntary acceptance of any risk and, in any case, s 149(3) of the Road Traffic Act 1988 precludes reliance on *volenti*: see, for example, *Pitts v Hunt* (1990).

Notes

Question 2

Gamma plc employs David as a driver and Elaine as a salesperson. One day, Elaine has to call on a customer but, as her car is being serviced, she asks David if he can drive her to the customer's premises. David agrees, but when they are in the car he tells Elaine that he must first call at his private house to collect a suit to take to the dry cleaners. Whilst on the way to David's house, David sees a patch of oil that has been spilt on the road and says to Elaine: '... see that oil – I'll show you how to control a skid.' David then drives onto the patch of oil, but fails to control the subsequent skid and hits a wall, injuring Elaine and damaging beyond repair a valuable painting that Elaine was carrying in her briefcase.

Advise Elaine.

Answer plan

This question involves the area of frolics and detours and the recent attitude of the courts when considering course of employment in situations involving deliberate acts by the employee, rather than negligent acts. As is typical of exam questions, however, an additional area is also tested, namely, the principle that a tortfeasor takes his victim as he finds him, and tests how that principle applies to property damage.

The following points need to be discussed:

- liability of David to Elaine;
- vicarious liability of Gamma plc for David's victim;
- consideration of course of employment as regards deliberate acts of the employee;
- possibility of *volenti*; and
- David takes Elaine as he finds her.

Answer

We must first decide whether Elaine can sue David and, if so, whether Gamma plc is vicariously liable for David's actions. Elaine must first show that David owes her a duty of care. In those situations where a duty of care has previously been found to exist, there is no need to apply the incremental formulation preferred by the House of Lords in *Caparo Industries plc v Dickman* (1990) or *Murphy v Brentwood District Council* (1990). We could note here the statement of Potts J at first instance in *B v Islington Health Authority* (1991), where he stated that in personal injury cases, the duty of care remains as it was pre-*Caparo*, namely, the foresight of a reasonable man (*Donoghue v Stevenson* (1932)), a finding that does not appear to have been disturbed on appeal (1992). In fact, a duty of care has been found to exist in a number of cases involving drivers and their passengers, for example, *Nettleship v Weston* (1971). However, even without knowledge of such cases, we could deduce the existence of a duty of care, as it is reasonably foreseeable that, by driving carelessly, a passenger may suffer injury.

Next, Elaine must show that David was in breach of his duty, that is, that a reasonable person or rather a reasonably competent driver in David's position would not have acted in this way (*Blyth v Birmingham Waterworks* (1856); *Nettleship v Weston* (1971)). It seems clear that a reasonable driver would not drive deliberately onto a patch of oil, and so David is in breach of his duty. Elaine will also have to show that this breach caused her injuries, and the 'but for' test in *Cork v Kirby MacLean* (1952) proves the required causal connection. Finally, Elaine will have to prove that the harm suffered was not too remote, that is, it was reasonably foreseeable (*The Wagon Mound (No 1)* (1961)). This should give rise to no problems, as all that Elaine will have to show is that some personal injury was foreseeable. Elaine will not have to show that the extent of the injury was foreseeable, or the exact manner in which the injury was caused (*Smith v Leech Brain* (1962); *Hughes v Lord Advocate* (1963)). Elaine could therefore sue David.

Next we must consider whether Gamma plc is liable for David's actions. We are told that the employer/employee relationship exists, and it seems clear that at the start of the journey, David is acting within the course of his employment. We need to consider, however, whether by calling at his house David has moved outside the course of his employment, that is, whether he is on a 'frolic of his own'. In *Whatman v Pearson* (1868), a driver who went home for lunch, contrary to his employer's instructions, was held to be still within the course of his employment. However, in *Storey v Ashton* (1869), employers were held not liable when the employee, after completing his work, embarked on a detour. It was held that this detour constituted a new and independent journey which had nothing to do with his employment and was, therefore, outside the course of his employment. This problem was considered by the House of Lords in *Williams v Hemphill Ltd* (1966), where a driver carrying some children undertook a considerable detour. The House held that the driver was still within the course of his employment, however. Lord Pearce stated that it was a question of fact in each case whether the deviation was so unconnected with the employer's business that the employee was on a 'frolic of his own' and, in *Williams*, the presence of the boys on the bus showed that it was not a frolic of the driver's own. Lord Pearce stated that had the driver in *Storey* been carrying some property of his employers, for instance, he might have remained in the course of his employment. Having considered *Joel v Morrison* and *Storey*, Lord Pearce stated that to constitute a frolic of his own the journey had to be entirely unconnected with the employer's business, as opposed to a mere detour for the employee's selfish purposes. However, on the facts of *Williams*, Lord Pearce held that the presence of passengers who the employee had to take to their destination made it impossible to say that the detour was entirely for the employee's purposes. Applying this criterion to our case, the presence of Elaine will make it impossible to say that the detour was undertaken entirely for David's selfish purposes and thus David remains in the course of his employment.

The next question we must consider is whether David remains in the course of his employment when he drives onto the patch of oil. The court has taken a much more restrictive approach to course of employment where 'deliberate' wrongful acts have occurred, as can be seen by decisions of the Court of Appeal in *Heasmans v Clarity Cleaning Co Ltd* (1987) and *Irving v Post Office* (1987). In *Heasmans*, an employer was held not to be vicariously liable for the actions of an employee who was employed to clean telephones, but who made unauthorised telephone calls costing some £1,400. The court noted that the employee was employed to clean the telephones and that in using them he had not cleaned them in an unauthorised manner, but had done an unauthorised act which had taken him outside the course of his employment. In *Irving*, the employee, who worked for the Post Office and was employed to sort mail, wrote some racial abuse concerning the plaintiff upon a letter addressed to the plaintiff. The employee was authorised to write upon letters, but only for the purposes of ensuring that the mail was properly dealt with. It was held that the employers were not vicariously liable for the actions of the employee, as in writing racial abuse he was doing an unauthorised act, and not an authorised act in an unauthorised manner. The court stated, *per* Fox LJ, that limits had to be set to the doctrine of vicarious liability, particularly where it was sought to make employers liable for the 'wilful wrongdoing' of an employee. Thus, in *General Engineering Services v Kingston and St Andrews Corp* (1989), the firemen who drove very slowly to a fire were held not to be in the course of their employment in so doing. The firemen were employed to travel to the scene of fire as quickly as reasonably possible, and in travelling slowly they were not doing an authorised act in an unauthorised manner, but an unauthorised act. Given the attitude of the courts to deliberate acts, it is submitted that in carrying out this deliberate act, David has moved outside the course of his employment and that Gamma plc will not be liable for his action. The fact that in driving onto the oil David was not acting for the benefit of his employer is not necessarily relevant to taking the act outside the course of his employment (*Lloyd v Grace, Smith and Co* (1912); *Century Insurance v Northern Ireland Road Transport Board* (1942)).

In *Lister v Hesley Hall Ltd* (2001), the House of Lords, in reviewing the law on course of employment, held that an employer will also be liable for unauthorised acts that are so connected with authorised acts that they may be regarded as modes, albeit improper modes, of doing authorised acts. Their Lordships held that it was necessary to concentrate on the relative closeness of the connection between the tort and the nature of the employment, taking a broad approach to nature of employment. The close connection approach was also used by the House of Lords in *Dubai Aluminium Co v Salaam* (2003) and the Court of Appeal in *Mattis v Pollock* (2004). Although *Lister* may seem at first glance to suggest that David is acting in the course of his employment when driving onto the patch of oil, it should be remembered that in *Lister* there was a very close connection between the tort and the course of employment. *Lister* is therefore not a general authority for widening the course of employment to cover deliberate acts, and it is submitted that a court would follow *Heasmans* or *Irving* in David's case.

David may seek to raise the *volenti* defence against Elaine. To do this successfully, David will have to show that Elaine voluntarily assented to the risk of damage, which seems unlikely (*Dann v Hamilton* (1939)). Although the defence of *volenti* succeeded in *Morris v Murray* (1990), this was a case where the risk was glaringly obvious from the outset. In any event, s 149(3) of the Road Traffic Act 1988 precludes reliance on *volenti* in road traffic situations (*Pitts v Hunt* (1990)).

As regards the damage to the valuable painting, David (and Gamma plc if he is still within the course of his employment) will be liable. David may not have foreseen that Elaine would be carrying such valuable property, but he could have foreseen that Elaine would be carrying some property and this will be sufficient (*Vacwell Engineering v BDH Chemicals* (1971)). When considering damage to property, a narrower attitude to foreseeability is taken than with harm to the person (*The Wagon Mound (No 1)* (1961)), and the egg shell skull rule in *Dulieu v White* (1901) cannot be applied without great caution. However, Elaine should be able to recover for the damage to her painting on the authority of *Vacwell*.

Notes

Question 3

Delta plc owns a small office. Delta asks Frank, an electrician, to undertake various works at the company premises and tells Frank that his work will take about one week.

While Frank is working in Delta's offices, he carelessly rewires a switch and Gloria, an employee of Delta, is injured when she uses the switch. Henry, who is visiting Delta in an attempt to sell some office equipment, is also injured when he trips over a length of electrical cable that Frank has left in a corridor.

Advise Gloria and Henry.

Answer plan

This question involves a consideration of the liability of an employer for the acts of an independent contractor. As such, it is less run of the mill than the standard vicarious liability questions and the answer will be much shorter, providing the student is aware of the relevant legal principles.

The following points need to be discussed:

- differentiation between employees and independent contractors;
- liability of employer for an independent contractor;
- non-delegable duties of employer; and
- liability of employer for acts of collateral negligence.

Answer

We must decide whether Gloria and Henry could sue Frank and, if so, whether Delta plc is liable for Frank's actions. Gloria must first show that Frank owes her a duty of care. In those situations where a duty of care has previously been found to exist, there is no need to apply the incremental formulation preferred by the House of Lords in *Caparo Industries plc v Dickman* (1990) or *Murphy v Brentwood District Council* (1990). We could note here the statement of Potts J at first instance in *B v Islington Health Authority* (1991), where he stated that in personal injury cases the duty of care remains as it was pre-*Caparo*, namely, the foresight of a reasonable man (*Donoghue v Stevenson* (1932)), a finding that does not appear to have been disturbed on appeal (1992). In fact, a duty of care has been found to exist in a similar situation in *Green v Fibreglass Ltd* (1958). However, even without knowledge of this case, we could deduce the existence of a duty of care as it is reasonably foreseeable that, by carelessly rewiring the switch, a person who subsequently uses it may suffer injury.

Next, Gloria must show that Frank was in breach of this duty, that is, that a reasonable person, or rather a reasonably competent electrician, in Frank's position, would not have acted in this way (*Blyth v Birmingham Waterworks* (1856); *Bolam v Friern Hospital Management Committee* (1957)). It seems clear that a reasonable electrician would not rewire a switch carelessly, and so Frank is in breach of his duty. Gloria will also have to show that this breach caused her injuries, and the 'but for' test in *Cork v Kirby MacLean* (1952) proves the required causal connection. Finally, Gloria will have to prove that the damage she has suffered was not too remote, that is, that it was reasonably foreseeable (*The Wagon Mound (No 1)* (1961)). This should give rise to no problems, as all that Gloria will have to show is that some personal injury was foreseeable; she will not have to show that the extent was foreseeable, or the exact manner in which the injury was caused (*Smith v Leech Brain* (1962); *Hughes v Lord Advocate* (1963)).

Given then that Frank is negligent, can Delta plc be held liable for his negligence? It seems clear that Frank was acting within the course of his employment. Provided that the relevant act (that is, the rewiring of the switch) was an authorised act, the fact that Frank has carried it out in a wrongful and unauthorised manner will not take the act outside the course of his employment (*Century Insurance v Northern Ireland Road Transport Board* (1942)). Consequently, we need to decide whether Frank is an employee of Delta plc or whether he is an independent contractor. It is extremely difficult to formulate a universal test for an employee. The original test, laid down in *Yewens v Noakes* (1880), was the 'control' test: the employer had the right of control as to the way in which the employee carried out his work. This test has obvious problems when the person concerned has a particular skill, especially if the employer himself does not possess that skill. In

Stevenson, Jordan and Harrison Ltd v MacDonald (1952), Denning LJ (as he then was) proposed the 'business integration' test: does the person do his work as an 'integral part of the business', when he will be an employee, or is he merely 'accessory' to it, when he will be an independent contractor? However, the test is just as difficult to apply as the control test and, in *Market Investigations v Minister of Social Security* (1969), it was held that a person was not an employee because she was not in business on her own account. This test also seems vague, although in *Andrews v King* (1991) it was described by Browne-Wilkinson VC as the 'fundamental test'. The modern approach of the courts is to eschew any single test and examine all the facts of the case. Looking at the facts of Frank and Delta plc, it would seem that Delta plc has the right to tell Frank what work is to be done but not how to do it, so that Frank is an independent contractor. In *Ready Mixed Concrete v Minister of Pensions* (1968), MacKenna J laid down several conditions for a contract of employment, one of which was that the worker agrees to be directed as to the mode of carrying out the work. MacKenna J held that this right of control was a necessary (though not sufficient) condition of an employment contract, and it was said to be particularly important in cases of temporary employment in *Interlink Express Parcels Ltd v Night Trunkers Ltd* (2001). However, this condition seems to be lacking in Frank's case, leaving him as an independent contractor. Thus, *prima facie*, Delta plc is not liable for Frank's actions.

However, as regards Gloria, Delta plc, as Gloria's employers, are under a non-delegable duty to take reasonable care for the safety of their employees (*Wilsons and Clyde Coal v English* (1938)). By the phrase 'non-delegable' one does not mean that the employer cannot delegate *performance* to an independent contractor, but that the employer cannot delegate *responsibility* for performance. Thus, although the standard rule is that an employer is not liable for the actions of an independent contractor (*Morgan v Girls Friendly Society* (1936); *D and F Estates v Church Commissioners* (1989)), Delta plc will be liable to Gloria for failing to take reasonable care for her safety.

As regards Henry, as he is not an employee of Delta plc, he cannot claim on this basis.

However, there are some situations where an employer is liable for the torts of an independent contractor. The employer is so liable in the following situations: if he has authorised the independent contractor to commit the tort (*Ellis v Sheffield Gas Consumers Co* (1853)); if he is negligent in choosing an independent contractor who is not competent (*Pinn v Rew* (1916)); and if a non-delegable duty is imposed upon him by common law, that is, a duty the performance of which can be delegated, but not the responsibility. Neither of the first two situations are relevant here, but one of the situations that may be relevant as regards common law non-delegable duties is where the independent contractor is employed to carry out work that is extra-hazardous (*Honeywill and Stein Ltd v Larkin Bros* (1934); *Alcock v Wraith* (1991)). In *Alcock*, the Court of Appeal held that a crucial question was: 'did the work involve some special risk or was it from its very nature likely to cause damage?' It is submitted that the work in question was not extra-hazardous, and so Delta plc will not be liable (*Salsbury v Woodland* (1969)).

If it were decided that Frank's work was extra-hazardous, Delta plc might be able to claim that Frank's negligence was merely collateral to the performance of his work and that they were not liable (*Padbury v Holliday and Greenwood* (1912)). Whether the negligence was in fact collateral would depend on the particular facts of the case, and we are not given sufficient details to come to any firm conclusion.

If Henry were to sue Delta plc under the Occupiers' Liability Act 1957, he would be met with the defence in s 2(4)(b) that Delta plc acted reasonably in entrusting the work to an independent contractor, and took such steps as were reasonable to satisfy themselves that the contractor was competent and that the work had been properly done. In *Haseldine v Daw* (1941), it was held that there was no need to check the work of contractors employed to repair a lift, as the work was technical. In *Woodward v Mayor of Hastings* (1945), it was held that as the work was non-technical, there was a need to check that it had been properly done. In Henry's situation, although the wiring of a switch is a technical task, Henry's injury arose because Frank left some electrical cable on the floor, which is not a technical matter. This situation seems closer to the carelessly cleaned step in *Woodward* than the carelessly repaired lift in *Haseldine*, so it is submitted that Henry could successfully sue under the Occupiers' Liability Act 1957.

Notes

2 Negligence – Duty of Care Generally and Restricted Situations

Introduction

Questions on the imposition of a duty of care usually take the form of an essay, typically on the development of a test for imposing a duty of care. It is also vital to be thoroughly familiar with situations in which limits are placed on the duty of care, that is, in particular, with the areas of negligent misstatement, nervous shock and economic loss. These topics, especially the first two, have been the subject of important decisions recently, and are popular with examiners either as problem or essay questions. It is vital to be aware of the recent leading cases in these areas.

Checklist

Students must be familiar with the following areas:

(a) development of a test for ascertaining the existence of duty of care;
(b) negligent misstatement:
- statements made to a known recipient and the special relationship;
- statements put into general circulation;

(c) nervous shock:
- criteria for recovery;
- restrictions on recovery;
- possible extensions of persons owed a duty of care; and

(d) economic loss:
- decision in *Junior Books v Veitchi* (1983);
- judicial retreat from *Junior Books*;
- current position regarding economic loss.

Question 4

Although the decision of the House of Lords in *Anns v Merton London Borough Council* (1978) was welcomed as a rationalisation of the law, it is now regarded as too simplistic, and the so called incremental approach is now universally used to determine the existence of a duty of care. Discuss this statement.

Answer plan

This is a typical essay question on the development of the modern test for imposing a duty of care.

The following points need to be discussed:

- brief background to *Anns v Merton London Borough Council* (1978);
- the *Anns* test;
- judicial retreat from *Anns*; and
- current approach of the courts, that is, the incremental approach.

Answer

Although an attempt to formulate a general test or principle to decide whether, in any particular circumstances, a duty of care arose was made in *Heaven v Pender* (1883), it was not until 1932 and the judgment of Lord Atkin in *Donoghue v Stevenson* (1932) that a general principle was firmly established. There were, of course, many situations in which the courts had recognised the existence of a duty of care, but no general principle existed to decide, in any new situations, whether a duty existed. The courts were for some time a little hesitant in applying the neighbour test as a general principle, but in *Home Office v Dorset Yacht* (1970), Lord Reid stated that the neighbour test was a statement of principle and should be applied unless there was some reason for excluding it. Proceeding from this, Lord Wilberforce, in *Anns v Merton London Borough Council* (1978), stated that in order to establish whether a duty of care exists, the question is to be approached in two stages (the two tier test). First, is there a sufficient relationship of proximity of neighbourhood between the wrongdoer and the person who has suffered damage such that, in the reasonable contemplation of the former, carelessness on his part may cause damage to the latter, in which case a *prima facie* duty of care arises? Secondly, are there any considerations which ought to negate or reduce or limit the scope of the duty or the class of persons to whom it is owed or the damage to which a breach of it may give rise? Thus, *Anns* neatly rationalised the law regarding the imposition of a duty of care by essentially stating that *Donoghue* applied unless there was a legal reason to disapply or modify *Donoghue*.

However, the courts have gradually come to realise that the imposition of a duty of care involves more complex considerations as *Anns* was used by the courts to expand the area of duty of care. Thus, in *Junior Books v Veitchi* (1983), the House of Lords held that liability could arise in respect of economic loss; and in *McLoughlin v O'Brian* (1983), the House considered the scope of nervous shock. Also, in *Emeh v Kensington, Chelsea and Westminster Area Health Authority* (1985), the court was prepared to extend the range of persons to whom a duty of care was owed. However, from 1985 onwards, the courts have retreated from the broad general principle of *Anns*.

The starting point of this approach was *Peabody Donation Fund v Sir Lindsay Parkinson* (1985). Lord Keith stated that the *Anns* test was not of 'a definitive character', and that although a relationship of proximity must exist before a duty of care can arise, the existence of duty must depend on all the circumstances of the case and the court must consider whether it is just and reasonable to impose a duty. Further criticism of the two tier test is to be found in *Leigh and Sillivan v Aliakmon Shipping*, both in the Court of Appeal (1985) and the House of Lords (1988). Oliver LJ, in the Court of Appeal, stated that *Anns* did not establish a new test of duty of care applicable in all cases, nor did it enable the court to determine policy in each case. The fear is that the first tier is so easily satisfied that it leaves too much to the second tier, namely, policy. In the House of Lords, Lord Brandon adopted a similar approach and claimed that *Anns* had been decided in a novel fact situation and not in a factual situation where no duty had previously been held to exist. In the latter situation, *Anns* was not applicable and no duty of care was owed, which seems to be a pre-*Donoghue* approach, let alone pre-*Anns*. This attack was continued in *Curran v Northern Ireland Co-Ownership Housing Association* (1987) by Lord Bridge, who stated that *Anns* obscured both the distinction between misfeasance and non-feasance and between contract and tort. Lord Bridge also approved the judgment of Brennan J in the High Court of Australia in *Sutherland Shire Council v Heyman* (1985), where the judge analysed the two tier test and rejected it, holding that the test for the existence of a duty of care was more complex. Brennan J also commented that it was preferable to develop novel categories of negligence incrementally and by analogy with established categories, rather than by a massive extension of a *prima facie* duty of care restrained only by indefinable considerations which ought to negate, reduce or limit the scope of the duty or the class of persons to whom it is owed.

The two tier test was again criticised in *Yuen Kun-Yeu v AG of Hong Kong* (1988) by the Privy Council and by the House of Lords in *Hill v Chief Constable of West Yorkshire* (1989), *Caparo Industries plc v Dickman* (1990) and *Murphy v Brentwood District Council* (1990). Also, the House of Lords again expressed its preference for the incremental approach of Brennan J. Indeed, in *Ravenscroft v Rederiaktiebolaget Transatlantic* (1991), Ward J held at first instance that the two tier test in *Anns* had been overruled by *Murphy*. In strict legal terms, this appears incorrect, but as the House of Lords has twice expressed its preference for the incremental approach over the two tier test in *Anns*, it seems highly unlikely that the *Anns* test will be applied in the future. The incremental approach as expounded by the House of Lords in *Caparo* involves the consideration of three factors: the loss must be reasonably foreseeable; there must be a relationship of proximity between the claimant and the defendant; and it must be fair, just and reasonable to impose a duty of care.

The first factor merely states that harm must be reasonably foreseeable, the test being the foresight of a reasonable person in the position of the defendant. The second factor, the relationship of proximity, is not expressed in a way that is immediately clear. The phrase was used in the *Anns* test and was considered in *Yuen Kun-Yeu*, where Lord Keith stated that proximity was a composite test describing 'the whole concept of necessary relationship between plaintiff and defendant'. This factor of proximity would seem, in many circumstances, to be the policy tier of *Anns* expressed in another way. The third factor, that it must be fair, just and reasonable to impose a duty of care, also seems very similar to the policy test in *Anns*. Indeed, in *Marc Rich and Co AG v Bishop Rock Marine* (1995), Balcombe LJ doubted whether the words 'fair, just and reasonable' imposed any additional test to that of 'proximity' – a statement that was not disapproved of when the case was heard in the House of Lords. In *Caparo*, Lord Oliver stated that the above three factors overlap and are really three facets of the same thing; in *Marc Rich*, the Court of Appeal stated that the three factors were not to be treated as wholly separate and distinct requirements, but rather as convenient and helpful approaches to the pragmatic question as to whether a duty should be imposed in any given case. The Court of Appeal went on to state that to take this approach would resolve all or virtually all the conflicts among the authorities, and again this statement was not criticised when the case was heard in the House of Lords.

Thus, the simple two stage test in *Anns* has been replaced with a more complex three stage (or possibly one stage according to *Marc Rich*) test in which the policy aspect of the court's decision has been restated in terms of proximity and fair, just and reasonable. However, despite the replacement of the wide test in *Anns* by the narrower incremental test in *Caparo*, the courts are willing, where appropriate, to impose a duty of care in novel fact situations. Thus, a referee in a rugby match owes a duty of care to players to ensure that no dangerous play occurs, both in a colts rugby match (*Smolden v Whitworth* (1997)) and in an adults match, whether that match is amateur or professional (*Vowles v Evans* (2003)). In *Pearson v Lightning* (1998), a golfer was held to owe a duty of care to golfers playing at another hole on the same course. Furthermore, the police have been found to owe a duty of care to persons with suicidal tendencies, whether that person is of sound mind (*Reeves v Commissioner of Police of the Metropolis* (1998)) or of unsound mind (*Kirkham v Chief Constable of Greater Manchester* (1990)).

Notes

Question 5

Martin was on a train, reading a copy of the *Financial Times*. Norman, who was sitting next to him, asked Martin what his job was and Martin replied that he was a stockbroker. Norman then asked Martin for some advice on investment and Martin jokingly replied that publishing seemed to be a good area. As a result of this discussion, Norman invested his life savings in publishing shares. A few months later, the value of these shares fell dramatically and Norman lost all his money. Depressed at being penniless, Norman then committed suicide.

Advise Norman's widow, Olive, of any remedy she might have against Martin.

Answer plan

This is a typical 'one to one' negligent misstatement question that requires a discussion of *Hedley Byrne* and later relevant decisions.

The following points need to be discussed:

- duty of care between Martin and Norman;
- social occasions and *Chaudry v Prabhakar* (1989); and
- Martin's liability for loss of money and Norman's death.

Answer

In order to advise Olive, we must first decide whether or not Martin owed Norman a duty of care. The traditional approach to this question is to see whether a special relationship exists between Martin and Norman, as laid down in *Hedley Byrne v Heller* (1964). In *Hedley Byrne*, there were held to be three elements to special relationships.

First, the representor must possess a special skill. In *Mutual Life and Citizens Assurance v Evatt* (1971), the Privy Council held that liability would only arise when the statement was made in the course of a business. However, as *Evatt* is a decision of the Privy Council, it is only of persuasive authority, and the dissenting minority held that *prima facie* a duty was owed by anyone who took it upon himself to make representations knowing that another person will reasonably rely on those representations. This view has been followed by the Court of Appeal in *Esso Petroleum v Mardon* (1976) and *Howard Marine v Ogden* (1976). More recently, in *Gran Gelato v Richcliffe* (1992), it was held as unarguable that a vendor of premises did not owe a duty of care to a purchaser to take reasonable care when answering inquiries regarding the property. In *Spring v Guardian Assurance plc* (1994), Lord Goff, in whose analysis Lord Lowry concurred, stated that the reference to 'special skill' in *Hedley Byrne* should be understood in the broad sense and that it would include special knowledge. It is suggested that this is the approach that would be followed today and thus Martin would satisfy this element of the test.

Secondly, the representee must reasonably rely on the representations. In *Smith v Eric Bush* (1990), it was held to be reasonable for the purchaser of a modest house to rely on the survey carried out by the lender's surveyor and in *Edwards v Lee* (1991) it was held that it was reasonable for the recipient of a reference provided by a solicitor concerning a client to rely on that reference. In *Royal Bank Trust (Trinidad) v Pampellonne* (1987), the Privy Council held that there was a difference between the giving of advice and the passing on of information, and that it may be more reasonable to rely on the former than the latter. Thus, the question here is whether it was reasonable for Norman to rely on Martin's statement. Here it does not seem reasonable for Norman to rely on Martin's reply to his question as to how to invest his life savings. In *Chaudry v Prabhakar* (1989), May LJ stated that a duty of care would not be imposed regarding statements made on social occasions, and it seems that the meeting between Martin and Norman could be described as such. Overall, therefore, it seems that Martin does not satisfy this requirement.

Thirdly, the representor must have some knowledge of the type of transaction envisaged by the representee, and although this does seem to be so from the facts of the problem, Martin owes no duty of care to Norman because the essential ingredient of reasonable reliance is absent.

In *Caparo Industries plc v Dickman* (1990), Lord Oliver analysed *Hedley Byrne* and held that the required relationship between the representor and the representee may typically be held to exist where:

(a) the advice is required for a purpose, whether particularly specified or generally described, which is made known, either actually or inferentially, to the representor when the advice is given;

(b) the representor knows, either actually or inferentially, that his advice will be communicated to the representee, either specifically or as a member of an ascertainable class, in order that it should be used by the representee for that purpose;

(c) it is known, either actually or inferentially, that the advice so communicated is likely to be acted upon by the representee for that purpose without independent inquiry; and

(d) it is so acted upon by the representee to his detriment.

Lord Oliver emphasised that these conditions were neither conclusive nor exclusive, but merely that the decision in *Hedley Byrne* did not warrant any broader propositions. Considering these conditions, it seems that Martin satisfies (a), (b) and (d), but that condition (c) is not satisfied and so again we conclude that Martin owes Norman no duty of care as regards the statement concerning shares. In *Hedley Byrne*, it was stated that a duty of care will arise where the representor voluntarily assumed a responsibility to the representee and it seems from the facts that Martin has not done this, as we are told he 'jokingly' recommended publishing shares. Despite the criticism of the assumption of responsibility concept by the House of Lords in *Caparo* and *Smith v Eric Bush*, more recent decisions of their Lordships have emphasised the importance of it: *Spring*; *Henderson v Merrett Syndicates* (1994); *White v Jones* (1995) – see especially the speeches of Lord Goff and the decision of the Court of Appeal in *Lennon v Metropolitan Police Commissioner* (2004). This is another factor which mitigates against the imposition of a duty of care on Martin. Thus, Olive cannot sue Martin in respect of the loss in value of the shares purchased by Norman, nor can she sue in respect of Norman's subsequent suicide. Even if Martin owed Norman a duty of care as regards the statement and was in breach of that duty, he would not be liable for Norman's subsequent suicide, as this is not a reasonably foreseeable consequence of the breach as is required by *The Wagon Mound (No 1)* (1961).

Norman's suicide could also be regarded as a *novus actus interveniens* which broke the chain of causation between the earlier negligence (if any) of Martin and the damage suffered by Norman. The criterion used by the courts seems to be whether the latter conduct by the claimant is reasonable or not. Thus, in *McKew v Holland and Hannen and Cubitts* (1969), the plaintiff was injured due to the negligence of the defendants and as a result suffered a residual intermittent loss of control of one leg. Despite this, the plaintiff went down a flight of steep stairs that had no handrail and while doing so his leg gave way and he was injured. It was held by the House of Lords that his action in going down a steep flight of stairs without a handrail was so unreasonable that it broke the chain of causation. Lord Reid stated that 'if the injured man acts unreasonably, he cannot hold the defendant liable for injury caused by his own unreasonable conduct'. In contrast, in *Wieland v Cyril Lord Carpets* (1969), the plaintiff, due to the original negligence of the defendants, experienced difficulty in using her bi-focal spectacles. She nevertheless continued to use them and as a result she too fell down a flight of stairs. It was held that the defendants were liable for this injury as the plaintiff had not been unreasonable in continuing to wear bi-focals. As by its nature suicide is unreasonable conduct, it is submitted that even if Martin were held to owe a duty of care in respect of his advice and to be in breach of that duty, he would not be liable for Norman's suicide, as that act would have broken the chain of causation. This is fortunate for Martin, because Norman's suicide would not allow the defence of *ex turpi causa non oritur actio* (*Kirkham v Chief Constable Greater Manchester* (1990)), nor would it allow the *volenti* defence, as presumably Norman was of unsound mind when he committed suicide (*Kirkham*).

Notes

Question 6

Neil, who is an accountant, writes a book entitled *How to Make a Fortune on the Stock Market*. Karen buys a copy from a bookshop and Peter is given a copy as a birthday present. Neil gives a copy to Rachel, his girlfriend, saying 'Have a look at this and see how clever I am', and a copy to Terence, his brother, saying 'Follow these tips and you will become a millionaire'. Karen, Peter, Rachel and Terence have followed the advice in the book and have lost a large amount of money, as the book is erroneous in several important aspects.

Advise Karen, Peter, Rachel and Terence.

Answer plan

This is a question on negligent misstatement that involves the 'one to one' situation, the placing of a statement into general circulation and a consideration of what constitutes a social occasion.

The following points need to be discussed:

- *Hedley Byrne v Heller* (1964) and Rachel and Terence;
- *Caparo Industries plc v Dickman* (1990) and Karen and Peter; and
- reasonable reliance and Rachel and Terence.

Answer

In advising the parties, we must consider whether Neil owes them a duty of care in respect of the statements that he has made in his book.

As regards Karen and Peter, Neil has put a statement into general circulation. In *Caparo Industries plc v Dickman* (1990), the House of Lords considered the situation of a person who places a statement into general circulation. This differs from those situations such as *Hedley Byrne v Heller* (1964) and *Smith v Eric Bush* (1990), where the representor knew that the advice was to be communicated to the representee, knew that it was very likely that the representee would rely on the advice and was fully aware of the nature of the transaction that the representee had in mind (the 'one to one' situation). The House in *Caparo* stated that the criteria for imposing a duty of care were foreseeability of damage, proximity of relationship and reasonableness or otherwise of imposing a duty.

In particular, where a statement put into more or less general circulation might be relied upon by strangers for one of a variety of different purposes, which the maker of the statement had no specific reason to anticipate, there was no relationship of proximity between the maker of the statement and any person relying on it, unless it could be shown that the maker knew that his statement would be communicated to the person relying on it, either as an individual or as a member of an identifiable class, specifically in connection with a transaction of a particular kind and this person would be very likely to rely on it in deciding whether to enter into that transaction.

Thus, it was held in *Al Saudi Banque v Clarke Pixley* (1989), that the auditors of a company owed no duty of care to a bank who lent money to the company, regardless of whether the bank was an existing creditor making further advances or was only a potential creditor. This is because in either case, even if it is foreseeable that the bank might request a copy of the company's accounts and rely on them, there was not a sufficiently close or direct relationship between the auditors and the bank to give rise to a degree of proximity necessary to establish a duty of care. In *Caparo*, it was held that the auditor of a public company owed no duty of care to a member of the public who relied on the accounts to buy shares in the company, because the court would not deduce a relationship of proximity between the auditor and a member of the public when to do so would give rise to unlimited liability on the part of the auditor. Furthermore, an auditor owed no duty of care to an individual shareholder who wished to buy more shares in the company, since an individual shareholder was in no better position than a member of the public. The auditors' statutory duty to prepare accounts was owed to the body of shareholders as a whole, the purpose being to enable the shareholders, as a body, to exercise informed control of the company and not to enable individual shareholders to buy shares with a view to profit.

In *James McNaughton Paper Group v Hicks Anderson* (1991), Neil LJ analysed the cases and identified a number of guidelines that may be relevant in determining whether the maker of a statement owed a duty of care to the recipient not to be negligent. Where a statement was acted on to the detriment of a recipient other than the person directly intended by the maker to act on it, consider the following:

(a) the purpose for which the statement was made. If the statement was made by the adviser for the express purpose of being communicated to the advisee, a duty of care may often arise. If the statement was made for a different purpose and for the benefit of someone other than the advisee, the precise purpose for which the statement was communicated must be carefully studied;

(b) the purpose for which the statement was communicated, for example, was the communication for information only or was it for some action to be taken?;

(c) the relationship between the adviser, the advisee and any relevant third party. If the statement was made for the benefit of someone other than the advisee, the relationship between the parties should be considered, for example, is the advisee likely to look to the third party (and through him to the adviser) for guidance?;

(d) the size of the class to which the advisee belongs. If the advisee is a single person or a member of a small class, it will be easier to infer that a duty of care was owed than if he was a member of a large class, especially where the statement was first made to someone outside that class;

(e) the state of knowledge of the adviser. This is a most important factor. Did the adviser know the purpose for which the statement was made and the purpose for which it was communicated? Any duty of care will be limited to the types of transactions of which the adviser had knowledge and will only apply where the adviser knows or ought to know that the statement will be relied on by a person or class of persons in connection with that transaction (*Caparo*). One should also consider whether the adviser knew that the advisee would rely on the statement without obtaining independent advice;

(f) reliance by the advisee. Was the advisee entitled to rely on the statement to take the action that he did? Did he in fact rely on the statement? Should he have used his own judgment? Should he have sought independent advice?

The most important guideline as regards Peter and Karen would appear to be (d), for although it might be argued that (a)–(c) and (e)–(f) are in favour of imposing a duty of care, the size of the class to which Karen and Peter belong is so great that there is insufficient proximity between them and Neil to impose a duty. It should also be noted here that in *Law Society v KPMG Peat Marwick* (2000), the Court of Appeal placed some importance on the proposition stated by Lord Oliver in *Caparo* that to rely on foreseeability alone would create a liability that was wholly indefinite in area, duration and amount, and would clearly not be just and reasonable.

In the light of these decisions, it would seem that although the criterion of foreseeability of damage can be met, there is insufficient proximity between Neil and Karen (see *Al Saudi Banque* and *Caparo*) and certainly insufficient proximity between Neil and Peter, as Peter was not even a purchaser of the book.

As regards Rachel and Terence, we have the one to one situation, so we need to consider whether the necessary special relationship exists between Neil and Rachel and between Neil and Terence. The usual approach to deciding whether a special relationship exists is to consider the three elements under *Hedley Byrne v Heller* (1964). Let us look at the position between Neil and Rachel first.

Under the first element, the representor must possess a special skill. In *Mutual Life and Citizens Assurance v Evatt* (1971), the Privy Council held that liability would only arise when the statement was made in the course of a business. However, as *Evatt* is a decision of the Privy Council, it is only of persuasive authority, and the dissenting minority held that *prima facie* a duty was owed by anyone who took it upon himself to make representations knowing that another person would reasonably rely on those representations. This view has been followed by the Court of Appeal in *Esso Petroleum v Mardon* (1976) and *Howard Marine v Ogden* (1976). Also, in *Gran Gelato v Richcliffe* (1992), it was held to be unarguable that a vendor of premises did not owe a duty of care to a purchaser to take reasonable care when answering enquiries regarding the property.

It is suggested that future courts would follow the approach of the Court of Appeal, and thus Neil would satisfy this element of the test.

Secondly, the representative must reasonably rely on the representations. In *Smith v Eric Bush* (1990), it was held to be reasonable for the purchaser of a modest house to rely on the survey carried out by the lender's surveyor. In *Edwards v Lee* (1991), it was held that it was reasonable for the recipient of a reference provided by a solicitor to a client to rely on that reference. In *Royal Bank Trust (Trinidad) v Pampellonne* (1987), the Privy Council held that there was a difference between the giving of advice and the passing on of information and that it may be more reasonable to rely on the former than the latter. Thus, the question here is whether it was reasonable for Rachel to rely on Neil's statement. On the facts, it does not seem reasonable for Rachel to rely on the statements in the book. In *Chaudry v Prabhakar* (1989), May LJ stated that a duty of care would not be imposed regarding statements made on social occasions, and it seems that the meeting between Neil and Rachel could be described as such. Overall, therefore, it seems that Neil does not satisfy this requirement.

Thirdly, the representor must have some knowledge of the type of transaction envisaged by the representee, and this does seem to be so from the facts of the problem.

Thus, on this analysis, Neil owes no duty of care to Rachel as, although he satisfies the first criterion, it is not reasonable for Rachel to rely on the statements in the book. Neil has given Rachel the book for a different purpose, namely to impress her, and has not suggested that she follows the advice contained therein. Additionally, Neil can argue that the advice was given on a purely social occasion, its function being to impress Rachel. Thus, it is submitted that in all these circumstances it is not reasonable for Rachel to rely on the contents of the book. Furthermore, as Neil has only given the book to Rachel to impress her, he will have no knowledge of the type of transaction undertaken by Rachel, as he does not intend her to act on the book in the first place. Hence, Neil can argue that he owes no duty of care to Rachel in respect of the advice in the book. In addition, Neil may avail himself of the defence that he has undertaken no voluntary assumption of responsibility to Rachel. In *Hedley Byrne*, it was stated that a duty of care would only arise where this criterion had been satisfied, and despite criticism of this concept by the House of Lords in *Caparo* and *Smith v Eric Bush*, more recent decisions of the House have emphasised the importance of this concept: *Spring v Guardian Assurance* (1994); *Henderson v Merrett Syndicates* (1994); *White v Jones* (1995) – see especially the speeches of Lord Goff. This is another factor which mitigates against the imposition of a duty of care on Neil vis à vis Rachel.

In *Caparo*, Lord Oliver analysed *Hedley Byrne* and held that the required relationship between the representor and the representee may typically be held to exist where:

(a) the advice is required for a purpose, whether particularly specified or generally described, which is made known, either actually or inferentially, to the representor when the advice is given;

(b) the representor knows, either actually or inferentially, that his advice will be communicated to the representee, either specifically or as a member of an ascertainable class, in order that it should be used by the representee for that purpose;

(c) it is known, either actually or inferentially, that the advice so communicated is likely to be acted upon by the representee for that purpose without independent inquiry; and

(d) it is so acted upon by the representee to his detriment.

Taking this approach, the required relationship will not arise because Neil is unaware that Rachel intends to rely on the advice in the book for the purposes of investing in the stock market.

However, when we come to Terence, it seems clear that Neil does owe Terence a duty of care, because he does satisfy the three criteria in *Hedley*. First, Neil takes it upon himself to make representations (*Esso Petroleum*; *Howard Marine*). Secondly, it is reasonable for Terence to rely on these representations as Neil has told him to do so. Finally, Neil must be aware of the transactions envisaged by Terence. Indeed it could be argued that by his actions and statement Neil has undertaken a voluntary assumption of responsibility to Terence – see *Spring, Henderson* and the recent case of *Lennon v Metropolitan Police Commissioner* (2004).

Thus, Neil owes a duty of care to Terence and is liable for the loss Terence has suffered.

Notes

Question 7

'Due to some of the difficulties of bringing an action in negligence, the tort of misfeasance in public office has recently become a popular alternative cause of action.'

Discuss the above statement with particular reference to the scope of the tort of misfeasance in public office.

Answer plan

This is a question on misfeasance in public office and its relation to negligence. This question should only be attempted by candidates who have a good, detailed grasp of the principles of misfeasance in public office.

The following points need to be discussed:

- outline of circumstances in which misfeasance in public office offers advantages over negligence; and
- discussion of the principles of misfeasance in public office with particular reference to *Three Rivers v Bank of England* (2000) and subsequent cases.

Answer

Claimants seeking to rely on the tort of negligence to recover for any loss they have suffered may find several difficulties in establishing their cause of action.

In particular they may find it difficult to establish the required degree of proximity between themselves and the defendant, which means that the plaintiff may be unable to establish the existence of a duty of care. Another problem might arise where the loss suffered is pure economic loss. Although recovery for pure economic loss was allowed in the House of Lords in *Junior Books v Veitchi* (1983), *Junior Books* has been so heavily criticised in later cases that it seems highly unlikely that the case will found any future actions involving pure economic loss. In such situations, the tort of misfeasance in public office may provide an alternative cause of action, and this tort may be appropriate where policy reasons restrict a duty of care in negligence.

Although the tort of misfeasance in public office can be found in the law reports in the 17th and 18th centuries, it was a comparatively unknown cause of action until 1985 and the case of *Borgion SA v Ministry of Agriculture* (1985). Here the plaintiffs, who were French turkey importers, had been banned by the defendants from exporting turkeys to England. The defendants admitted that the true purpose of this ban was the protection of British turkey products and that this constituted a breach of Art 30 of the Treaty of Rome. Nevertheless, the defendants claimed that they were not liable for misfeasance in public office as they had no intent to injure the plaintiffs, but rather had the intent to protect British interests. The Court of Appeal found that malice is not an essential ingredient of the tort: it was sufficient that the defendant knew that he had acted unlawfully and that his acts would injure the plaintiff. As in this case the plaintiff had suffered only pure economic loss, an action in negligence was almost certain to fail.

The next example of this tort concerned a local authority. In *Jones v Swansea City Council* (1991), the House of Lords held that the plaintiff could sue the council for misfeasance in public office if she could prove that the majority of councillors who had voted for a resolution had done so with the aim of damaging the plaintiff with knowledge of the unlawful nature of this act. Again in this case the plaintiff had suffered only pure economic loss.

The House of Lords was called upon to undertake a comprehensive review of the tort in *Three Rivers District Council v Bank of England (No 3)* (2000). Their Lordships held that the tort has the following ingredients:

(a) The defendant must be a public officer. This clearly covers government departments and local authorities. If a local authority exercises private law functions as, for example, a landlord, this would satisfy this ingredient.

(b) The exercise of power must be as a public officer.

(c) The state of mind of the defendants. A study of the case law shows two different forms of liability for the tort exist:

- cases where a public power is exercised for an improper purpose with the specific intention of injuring a person or persons (the targeted malice limb); and
- cases where a public officer acts in the knowledge that he had no power to do the act complained of and that it would probably injure the claimant (the illegality limb).

Both limbs involve bad faith. In the targeted malice limb, the bad faith is the exercise of public power for an improper purpose; in the illegality limb, the bad faith is the lack of honest belief on the part of the public officer that the act is lawful.

The House of Lords made it clear that for the illegality limb, reckless indifference as to the illegality and its probable consequences is sufficient to establish the required mental element. The recklessness must be subjective, so that the plaintiff must prove that the defendant lacked an honest belief in lawfulness of his actions or wilfully disregarded the risk of unlawfulness.

(d) Duty to the claimant. Although the Court of Appeal in *Three Rivers DC* held that proximity between the claimant and the defendant was required, this finding was expressly overturned by the House of Lords. Their Lordships held that the required mental element will keep the tort within reasonable bounds, and that there was no need to introduce proximity as a control mechanism.

(e) Causation is an essential ingredient of the tort.

(f) Damage covered includes pure economic loss, but the plaintiff must suffer special damage, ie, loss which is specific to the plaintiff and not suffered in common with the general public. However, in *Watkins v Secretary of State for the Home Department* (2004), the Court of Appeal held that proof of damage was not an essential ingredient of the tort. In *Watkins*, the claimant's right of access to the courts had been infringed, and the court stated that this was a right of such a level of importance that where it was maliciously infringed by prison officers, the tort was complete without proof of special damage. It could, of course, be argued that the loss of such an important constitutional right of itself constitutes special damage.

(g) The test for remoteness for this tort is not reasonable foreseeability but knowledge by the defendant that the decision would probably damage the plaintiff.

The *Three Rivers DC* case arose out of the collapse of a bank (BCCI). Many of the bank's depositors brought misfeasance proceedings against the Bank of England. The plaintiffs alleged senior officials at the Bank of England licensed BCCI when they knew that doing so was unlawful, had shut their eyes to activities at BCCI once the licence was granted and had failed to close down BCCI when they ought to have done so. Had this case been brought in negligence, the problem of pure economic loss would have caused insurmountable difficulties, and in addition the claimants could have faced problems in establishing the necessary degree of proximity to found a duty of care.

Some short time after *Three Rivers DC* was heard, the Court of Appeal were faced with a misfeasance case involving for the first time personal injury or death: *Akenzua v Secretary of State for the Home Department* (2003). Here an immigration officer attached to a special police unit allowed a gangster with a record of violent crime in Jamaica to remain in the UK illegally in return for information concerning criminal activity in the UK. The gangster sexually assaulted and killed a friend of the woman with whom he lived. The personal representatives of the deceased sued the Home Secretary and the Police in misfeasance in public office. It is highly likely that an action in negligence would have failed for insufficient proximity, and possibly on public policy grounds: see *Hill v Chief Constable of West Yorkshire* (1989). The case was brought under the illegality limb of the tort and the defendant claimed that the deceased was not a member of a closely defined class at risk from the gangster's presence in the UK. Essentially the court had to decide whether the claimant had to prove (i) that the probable harm was to the claimant or a class of which the claimant was a member or (ii) only that the probable harm was to someone and that someone turned out to be the claimant. The court held that the second requirement was the correct one, and that the first requirement amounted to an attempt to introduce proximity into the tort.

From the above discussion it can be seen that the tort of misfeasance in public office is particularly apposite where the claimant may face problems of proximity, pure economic loss or public policy considerations should the action be brought in negligence.

Notes

Question 8

Arthur, Basil and Charles are working together on an electrical installation in Edward's factory. Charles leaves the workplace to collect some equipment from the stores in the next building and, while he is at the stores, he hears an explosion coming from the workplace. The explosion kills Arthur. Debra, a factory first aid worker employed by Edward, goes to try to help Arthur. Arthur's body is identified at the morgue by George, his father, but his mother, Hilda, cannot bring herself to view the mutilated body.

The explosion was the responsibility of Edward.

Advise all the parties, who have suffered nervous shock, as to whether they can sue Edward.

Answer plan

This is a wide-ranging question on nervous shock and should only be attempted by candidates who have a good knowledge of recent developments in this area.

The following points need to be discussed:

- criteria for liability laid down in *Alcock v Chief Constable of South Yorkshire Police* (1991);
- effect of employer/employee relationship as stated in *Frost v Chief Constable of South Yorkshire Police* (1997); and
- status of rescuers as considered in *Frost*.

Answer

The law on nervous shock, or psychiatric damage as it is sometimes called, has developed considerably since the original refusal to impose liability in *Victorian Railway Commissioners v Coulthas* (1888). It has progressed from allowing recovery where the claimant is reasonably put in fear of his own safety (*Dulieu v White* (1901)) to allowing recovery for a wide range of persons. However, with the exception of rescuers (*Chadwick v British Transport Commission* (1967)), such persons have usually been close family members (see the speech of Lord Wilberforce in *McLoughlin v O'Brian* (1983)). However, all liability for nervous shock must now be considered in the light of two decisions in the House of Lords: *Alcock v Chief Constable of South Yorkshire Police* (1991) and *White v Chief Constable of South Yorkshire Police* (1999).

In *McLoughlin*, the House of Lords considered the area of nervous shock and held that the test to be applied was whether it was reasonably foreseeable that the plaintiff would suffer nervous shock as a result of the defendant's negligence. However, the House adopted two distinct approaches to liability. Lord Wilberforce held that, as nervous shock is capable of affecting such a wide range of persons, there was a need for the law to place some limitations on claims. He stated that there were three elements inherent in any claim, namely, the class of persons who could claim, the proximity of such persons to the accident in time and space, and the means by which the shock was caused. In contrast, Lord Bridge considered that this approach would place arbitrary limits on recovery and preferred the test of reasonable foreseeability *simpliciter*. In *Alcock*, the House of Lords adopted Lord Wilberforce's approach and held that a plaintiff could only recover for nervous shock if he satisfied both the test of reasonable foreseeability (that he would be affected because of the close relationship of love and affection with the primary victim) and the test of proximity to the tortfeasor (in terms of physical and temporal connection between the claimant and the accident).

Hence, a claimant could only recover if:

(a) his relationship to the primary victim was sufficiently close that it was reasonably foreseeable that he might suffer nervous shock;

(b) his proximity to the accident or its immediate aftermath was sufficiently close in both time and space; and

(c) he suffered nervous shock through seeing or hearing the accident or its immediate aftermath.

Thus, a claimant does not satisfy the tests of reasonable foreseeability or proximity unless the psychiatric illness was caused by sudden nervous shock through seeing or hearing the accident or its immediate aftermath. Also, a claimant who suffers nervous shock caused by being informed of the accident by a third party, does not satisfy these tests.

The House of Lords also held that the class of persons who may claim for nervous shock was not limited to particular relationships such as parent and child or husband and wife. The court went on to suggest that a bystander who witnessed a particularly horrific catastrophe might be able to recover. However, in *McFarlane v EE Caledonia Ltd* (1994), the Court of Appeal held that despite the *dicta* of three Law Lords in *Alcock*, a mere bystander or witness could not recover unless there was both sufficient proximity in time and space to the accident, and a close relationship of love and affection with the primary victim. To hold otherwise, held the court, would be to reduce the test for recovery to reasonable foreseeability, which goes against the whole judgment in *Alcock*.

Before we apply these criteria to the parties in question, it is worth noting that by nervous shock we mean actual mental injury or psychiatric illness, and that mere grief and sorrow are insufficient (*Brice v Brown* (1984)). However, in *Re The Herald of Free Enterprise* (1989), it was held that post-traumatic stress disorder and pathological grief in excess of normal grief are recognised psychiatric illnesses for which compensation can be awarded. Thus, the Court of Appeal held in *Nicholls v Rushton* (1992) that a plaintiff who had undergone no physical injury, but who suffered a nervous reaction falling short of an identifiable psychological illness, could not recover. Hence, all the potential claimants in this question would have to show that they had suffered an actual psychological illness before any recovery would be possible.

Basil

At first glance, Basil might appear to be a mere bystander at Arthur's fatal accident. His position is, however, somewhat more complex than this as he is a co-worker of Arthur. When *White* was heard in the Court of Appeal (under the name *Frost v Chief Constable of South Yorkshire Police* (1997)), it was held that an employer who negligently caused physical harm to an employee was liable to a fellow employee working on the same task who suffered nervous shock, either from fear for his own safety or through witnessing what happened, and that the *Alcock* criteria did not apply in such cases. Unfortunately for Basil, the House of Lords overruled the Court of Appeal on this point, so that Basil cannot simply rely on his employee/employer relationship with Edward to found liability.

Of course, if Basil was reasonably put in fear of his own safety by the explosion that killed Arthur, Basil could recover on the well-established authority of *Dulieu*.

Basil has some other routes to liability open to him. In *Page v Smith* (1995), the House of Lords held that in the case of nervous shock, it is essential to differentiate between primary and secondary victims. In claims by secondary victims, the *Alcock* criteria apply, but not in the case of primary victims. Thus, if Basil comes within the category of a primary victim, he will be able to recover. In view of Basil's spatial proximity to the accident, he would be treated as a primary victim; in *Young v Charles Church (Southern) Ltd* (1997), the plaintiff was treated as a primary victim because he was within the area of physical risk created by the defendant's negligence. As stated in *Page*, once the defendant is under a duty not to cause personal injury to the claimant, it is immaterial whether the injury caused was physical or psychiatric. Thus, unlike the police officers in *White*, Basil may be a primary victim as he was working together with Arthur when the explosion occurred.

Charles

Like Basil, Charles cannot rely on a simple employee/employer relationship with Edward to succeed in his case. Charles will find it difficult to establish the existence of a duty of care. As he is a secondary victim, not a primary victim, the *Alcock* criteria apply to him. He would appear not to satisfy the criterion of close proximity in space, although presumably he suffered nervous shock through the sight of the immediate aftermath of the accident. However, in *Hunter v British Coal Corp* (1998), a co-worker who was some 30 metres away from an accident failed to recover for nervous shock, and a similar decision was reached in *Duncan v British Coal Corp* (1997). Thus, it seems that Charles cannot establish a sufficient proximity in space to the accident to found a duty of care.

Debra

Although Debra also cannot claim the existence of a duty of care purely by virtue of the employer/employee relationship, she could attempt to recover on the grounds that she is a rescuer. Traditionally, the law has placed rescuers in a special category which is not subject to the *Alcock* criteria, as was recognised in *McLoughlin* and reiterated in *Alcock*. Thus, *prima facie*, Debra could rely on her status as a rescuer (*Chadwick v British Transport Commission* (1967)).

However, Debra should be advised that the topic of rescuers has been subject to a re-analysis in *White*. In *White*, the House of Lords held that only rescuers who exposed themselves to danger or who reasonably believed they were doing so could recover without meeting the *Alcock* criteria. Thus, in *Greatorex v Greatorex* (2000), the High Court, following *White*, refused to allow recovery to a rescuer *qua* rescuer as he had not been exposed to danger in the course of rescue, nor had he been in reasonable fear of such danger. Also, consider *Duncan*, where a fellow employee was crushed to death and the plaintiff arrived on the scene within four minutes and attempted unsuccessfully to revive the deceased. It was held that although the plaintiff had suffered a harrowing experience, what he had seen was an inanimate body which would not have affected a person of reasonable fortitude. Thus, the blanket imposition of a duty of care for rescuers no longer applies.

Applying this restatement of the law on rescuers to Debra, since Arthur was dead when Debra arrived on the scene, it would seem that she cannot recover.

George

George is of course a secondary victim and therefore he will have to satisfy the *Alcock* requirements.

As George is Arthur's father, the law will presume a close relationship of love and affection, but it is always open to Edward to disprove this if he can. The problem that George has is whether he is sufficiently close in time and space to the accident or its immediate aftermath. In *McLoughlin*, the plaintiff came to the aftermath of the accident within two hours of the accident, and Lord Wilberforce regarded this as being on the borderline of recovery. Thus, the time gap between the accident and George seeing Arthur's body will be of vital importance. George will also have to show that his nervous shock was caused through the sight of Arthur's mutilated body, that is, by the aftermath of the accident, rather than by his fear of what he might see.

As both Charles and George may have problems in establishing the necessary proximity to the accident or its immediate aftermath in time and space, they should both be advised that in *W v Essex County Council* (2000), the House of Lords seemed to take a very wide view of what constituted the required proximity. In this case, the claimant parents found out after the event that their children had been subjected to sexual abuse by a foster child placed with them by the defendant local authority. As a result, the parents suffered nervous shock, and the House of Lords refused to strike out their claim, stating that they were not convinced that the parents had to come across the abuse or abused immediately after the event or events had occurred. Although a full discussion of this point must await the full trial of *W v Essex County Council* (as the case was only reported as a striking out case), it might be that in certain circumstances the strict time requirements that have been set in *Alcock* and *McLoughlin* might be relaxed.

Hilda

Hilda, as Arthur's mother, has the required relationship of love and affection with Arthur. Again, the law will presume such a relationship, and it will be up to Edward to prove, if he can, that in this particular case the relationship does not exist. However, assuming that the relationship does exist, a problem arises, in that Hilda neither witnessed the accident nor its immediate aftermath, as she did not see Arthur's mutilated body. She thus fails to satisfy the third criterion in *Alcock*. Hilda should be advised that the High Court decision in *Ravenscroft v Rederiaktiebolaget* (1991), in which a mother was allowed to recover in respect of nervous shock caused by her son's death, even though she was not present at the accident or its immediate aftermath, was disapproved of in *Alcock* and has been overruled by the Court of Appeal (1992). Hence, Hilda will be unable to recover for her nervous shock.

Notes

Question 9

'The House of Lords has stated in the clearest possible terms in *White v Chief Constable of South Yorkshire Police* (1999) that the law on nervous shock or psychiatric damage is so illogical that only Parliament can come up with a solution.'

Discuss the above statement.

Answer plan

This essay on nervous shock requires the candidate not merely to recite the current state of the law, but also to highlight any inconsistencies that exist and to discuss how a statute might improve the situation.

The following points need to be discussed:

- criteria for liability in nervous shock, *per Alcock v Chief Constable of South Yorkshire* (1991);
- uncertainties as regards possible claimants – rescuers, intervention of third parties, lapse of time;
- extent to which a statute might improve the current situation; and
- possible problems that a statute might bring.

Answer

The law on nervous shock, or psychiatric damage as it is sometimes now called, has developed considerably since the original refusal to impose liability in *Victorian Railway Commissioners v Coulthas* (1888). It has progressed from allowing recovery where the claimant was reasonably put in fear of his own safety (*Dulieu v White* (1901)) to allowing recovery for a wide range of persons. However, with the exception of rescuers (*Chadwick v British Transport Commission* (1967)), these persons have usually been close family members of the victim (see the speech of Lord Wilberforce in *McLoughlin v O'Brian* (1983)). However, all liability for nervous shock must now be considered in the light of the decision of the House of Lords in *Alcock v Chief Constable of South Yorkshire* (1991).

In *McLoughlin*, the House of Lords considered the area of nervous shock and held that the test to be applied was whether it was reasonably foreseeable that the claimant would suffer from nervous shock as a result of the defendant's negligence. However, the House of Lords adopted two distinct approaches to liability. Lord Wilberforce held that as nervous shock was capable of affecting such a wide range of persons, there was a need for the law to place some limitation on claims. He considered that there were three elements inherent in any claim, namely, the class of persons who could claim, the proximity of such persons to the accident in time and space, and the means by which the shock was caused. Lord Bridge held that this approach would place arbitrary limits on recovery, and preferred the test of reasonable foreseeability *simpliciter*.

In *Alcock*, the House of Lords adopted Lord Wilberforce's approach and held that a claimant could only recover for nervous shock if he satisfied both the test of reasonable foreseeability that he would be so affected because of the close relationship of love and affection with the primary victim, and the test of proximity to the tortfeasor in terms of physical and temporal connection between the claimant and the accident.

Hence, a claimant could only recover if:

(a) his relationship to the primary victim was sufficiently close that it was reasonably foreseeable that he might suffer nervous shock;

(b) his proximity to the accident or its immediate aftermath was sufficiently close in both time and space; and

(c) he suffered nervous shock through seeing or hearing the accident or its immediate aftermath.

Thus, a claimant does not satisfy the tests of reasonable foreseeability or proximity unless the psychiatric illness was caused by sudden nervous shock through seeing or hearing the accident and its immediate aftermath. Also, a claimant who suffered nervous shock caused by being informed of the accident by a third party does not satisfy these tests.

Thus, given the television broadcasting guidelines which forbade the transmission of pictures of any identifiable individuals involved, persons who witnessed the disaster live on television had not suffered nervous shock induced by the sight and sound of the event, as they were not in proximity to the event and did not suffer shock in the sense of a sudden assault on the nervous system. The House of Lords also held that the class of persons who may claim for nervous shock was not limited to particular relationships such as husband and wife or parent and child, and went on to suggest that a bystander who witnessed a particularly horrific catastrophe might be able to recover and that, in certain circumstances, a claimant might recover on witnessing an event on contemporaneous television (for example, where the claimant knew that the primary victim would be injured in a live televised event, even though the primary victim was not identified in the televised pictures). However, in *McFarlane v EE Caledonia Ltd* (1994), the Court of Appeal held that despite the *dicta* of three Law Lords in *Alcock*, a mere bystander or witness of horrific events could not recover unless there was both sufficient proximity in time and space to the accident and a close relationship of love and affection with the primary victim. To hold otherwise, held the court, would reduce the test for recovery to pure reasonable foreseeability, which goes against the whole judgment in *Alcock*.

We should note here that by nervous shock the courts mean actual mental injury or psychiatric illness, and that mere grief and sorrow are insufficient (*Brice v Brown* (1984)). However, in *Re The Herald of Free Enterprise* (1989), it was held that post-traumatic stress disorder and pathological grief in excess of normal grief are recognised psychiatric illnesses for which compensation can be awarded. Thus, the Court of Appeal held in *Nicholls v Rushton* (1992) that a plaintiff who had undergone no physical injury, but suffered a nervous reaction falling short of an identifiable psychiatric illness, could not recover.

The courts have also allowed recovery where property damage has occurred. Thus, in *Attia v British Gas* (1988), a plaintiff was allowed to recover when she saw her house being burnt to the ground. Presumably, since *Alcock*, a claimant will have to show that the property was of such a nature that if a claimant witnessed its destruction, it was reasonably foreseeable that nervous shock would follow, as well as satisfying the criteria of proximity and of seeing the accident through his own senses.

At first glance, in *Alcock*, the House of Lords seemed to have widened considerably the range of potential claimants, although a close reading of the judgments might suggest that this range has been narrowed in some circumstances. Thus, their Lordships rejected the concept of limiting the class of persons who can claim to specified relationships such as spouses or parents and children in favour of the close relationship test. This is both logical and just in that, *per* Lord Keith, it is the existence of the close tie of love and affection which leads to nervous shock. Thus, the spouse or parent will be presumed to have such close ties of love and affection, and siblings and other relatives will have to prove such ties. Presumably, it would be open to the defendant in the cases of spouses to rebut the presumption by proving (say) that the partners have separated and have not been living together for some years. This wide approach, however, is not free from difficulties. It seems that *Alcock* would allow recovery by a particularly close friend who can satisfy the criteria of love and affection, but how is a defendant to reasonably foresee the existence of such a close friend? While it is foreseeable that the primary victim of an accident may have a spouse or children or a brother or sister, is the existence of such a friend reasonably foreseeable? Given the readiness of some judges to foresee a great deal and others to take a narrower view, can this approach be said to bring certainty or logic to the law? It may be just from the point of view of the secondary victim, but is it just as regards the defendant to impose such wide liability? Another area that gives rise to problems of justice and uncertainty arises from the second requirement that the claimant's proximity to the accident or its immediate aftermath is close in both time and space. The necessity for such a requirement is obvious, in that the claimant should not be allowed to claim a long time after the accident, but just what is meant by being close in both time and place? In *Alcock*, Lord Ackner was not prepared to allow recovery to a plaintiff who saw the body of a brother-in-law at the mortuary some eight hours after the accident, and Lord Wilberforce stated in *McLoughlin* that a two hour delay period was at the margin of the time span for recovery. This seems to be an arbitrary timescale which would appear to suggest that a claimant who is contacted by mobile telephone and told to attend at a hospital and has a Porsche or Ferrari may be able to recover, whereas a claimant who has to depend on public transport may not. Is a claimant who is away on business and on return identifies a dead spouse any different from a person who is called from work to identify a dead spouse?

The problem as to just what is meant by the claimant's proximity to the event or its immediate aftermath has been rendered even more confusing by the decision of the House of Lords in *W v Essex County Council* (2000). Here the claimant parents fostered a youth placed with them by the defendant local authority. The youth committed severe acts of sexual abuse on the claimants' children, and when the parents discovered what had happened they suffered psychiatric illness. The defendant authority sought to strike out the claim, but this was refused by the House of Lords. Lord Slynn, with whom all the other Law Lords agreed, stated that it was by no means certain that the parents would fail to satisfy the required proximity to the event or its immediate aftermath in both time and space. Lord Slynn stated that he was not certain that in this case the parents would have to come across the abused or abuser immediately after the sexual event. Given the statements regarding time in *Alcock* and *McLoughlin* above, this seems a very strange proposition of law. In addition, the parents were not witnesses to the abuse – they found out about it some time later. Their position is analogous to that of parents who are told of their offspring's involvement in an accident after the event, and in *Ravenscroft v Rederiaktiebolaget Transatlantic* (1992), such a person was denied recovery for nervous shock. Hence, there would appear to be considerable uncertainty in deciding in any particular case whether or not there was sufficient proximity to the accident or its immediate aftermath in time and space.

The word 'shock' in nervous shock is also not without its problems. In *Walters v North Glamorgan NHS Trust* (2003), the Court of Appeal had to decide whether an event which lasted for 36 hours could constitute 'shock', or whether it was a gradual assault of the mind over a period of time. On the facts, the court held that this 36 hour period constituted an event, and stated that the present law permits a realistic view to be taken in each individual case. As in similar cases, for example, *Taylorson v Shieldness Produce Ltd* (1994), where it has been held that there was a dawning realisation rather than a sudden shock over such a period of time, it is clear that the facts and medical reports regarding just how the psychiatric damage eventuated are of the utmost importance.

Finally, *Alcock* retained the rule that the nervous shock must be caused through seeing or hearing the accident or its immediate aftermath. Thus, if a mother attends a hospital to be told that her children have been burnt to death and feels unable to see the bodies, but still suffers nervous shock, she cannot recover. Presumably, if she did see the bodies, it would be open to the defendant to argue that it was the news of the death of her children, related to her by a nurse or doctor, that caused the shock, rather than the sight of the bodies. This seems to be a most illogical and unjust result, but it follows from *Alcock*.

Two recent decisions of the House of Lords have attempted to introduce some logic into the area of nervous shock. In *Page v Smith* (1995), their Lordships held that once it can be established that a defendant is under a duty of care to avoid causing personal injury to a claimant, it is immaterial whether the injury caused is physical or psychiatric. Thus, providing that it is reasonably foreseeable that the claimant might suffer personal injury, that will suffice in a nervous shock claim. The House went on to state that in nervous shock cases, it is vital to distinguish between primary and secondary victims, as only secondary victims are subject to the restrictions in *Alcock*. Thus, for primary victims, the illogical distinction between physical and psychiatric injury has been abolished.

In *White v Chief Constable of South Yorkshire Police* (1999), the House of Lords removed what most commentators had recognised was an illogical and unjust distinction between claimants that had been brought about by the decision of the Court of Appeal in this case (reported as *Frost v Chief Constable of South Yorkshire Police* (1997)). The case concerned the Hillsborough disaster and, while in *Alcock* the claims of the deceased's families were not allowed, in *Frost* the claims of the police officers who were present were (mostly) allowed. The Court of Appeal reached this decision by holding that as the plaintiff police officers were in an employee/employer relationship with the defendant Chief Constable, a duty of care was owed to them where injury was caused by the negligence of the Chief Constable. Thus, the distinction between primary and secondary victims was irrelevant in the employment situation. The Court of Appeal also held that the police officers were rescuers and could recover relying on that status, which also did not involve the application of the *Alcock* criteria. This decision was overturned by the House of Lords, which held that an employee who suffered psychiatric injury in the course of employment had to prove liability under the general rules of negligence, that is, employers' liability is not a separate tort with its own rules, but merely an aspect of the law of negligence. Their Lordships also went on to deal with the rescuer argument, and held that a rescuer had to show that he had exposed himself to danger or reasonably believed he was so doing. Thus, rescuers are not to be treated as primary victims merely because they are rescuers. Consequently, in *Greatorex v Greatorex* (2000), the High Court, following *White*, refused to allow recovery to a rescuer *qua* rescuer as he had not been exposed to danger in the course of the rescue, nor had he been in reasonable fear of such danger.

While *White* brings some logic to the area of nervous shock, in that employees are treated in an identical manner to other claimants, it has weakened the position of rescuers. Furthermore, because of the rule in *Ogwo v Taylor* (1987), professional rescuers are treated in exactly the same way as pure volunteer rescuers, which might seem illogical.

In *Hunter v British Coal Corp* (1998), the Court of Appeal attempted to formulate some logical guidelines to help distinguish between a participant and a bystander, the *Alcock* criteria being applicable only to the latter. The Court of Appeal held that a claimant is a participant if he reasonably believes he is in physical danger as a result of the accident, or if he is an unwitting instrument of another person's negligence and therefore feels responsible for the accident. The claimant in *Hunter* was 30 metres away when the accident occurred and never returned to the scene of the accident. Thus, he was not a participant in the accident and his claim failed because he could not satisfy the *Alcock* criteria. A similar decision was reached by the Court of Appeal in *Duncan v British Coal Corp* (1997), where the plaintiff was 300 metres away from the accident.

A further degree of uncertainty arises in this area as regards those statements of the law lords in *Alcock* that may be regarded as *obiter*, rather than forming part of the *ratio decidendi*. In *Alcock*, three law lords recognised the possibility of a mere bystander recovering after witnessing a particularly horrific accident, but in *McFarlane v EE Caledonia Ltd* (1994), the Court of Appeal held that such a bystander could not recover unless a close relationship of love and affection existed between him and the primary victim. To allow recovery in such a case would be to reduce the criteria to pure reasonable foreseeability, which runs counter to the whole of the judgment in *Alcock*.

Finally, it should be noted that when the High Court had to consider an extension to the law of nervous shock in *Greatorex*, where the rescuer was the father and the personal injury to the primary victim was self-inflicted due to the primary victim's own negligence, the court relied almost entirely on policy considerations in denying recovery to the rescuer. As decisions which involve matters of policy are notoriously difficult to predict and are subject to a wide amount of judicial variation, this adds to the uncertainty prevalent in the area of nervous shock.

Thus, it can be seen that the current state of the law on nervous shock is illogical and uncertain in some respects. The enactment of a statute could remove some of the uncertainty, but whether this would be at the expense of justice and flexibility is a problem. Should the law specify categories of relationship into which a claimant must fit to recover? Should the criteria for proximity in time and space be defined? Surely, the only limits that could be so defined are 'reasonable' proximity in time and space, which are hardly certain. A statute could remove the necessity for direct sight or sound of the accident or its immediate aftermath, and allow recovery where the claimant is informed by a third person, subject to the claimant proving that it was the accident that caused the nervous shock, rather than his mind imagining what the accident and its effects were. A statute would not at a stroke solve all the problems associated with nervous shock, but it could introduce a welcome degree of certainty and logic into this area of the law.

Notes

Question 10

When one considers the tortious liability of the emergency services, it becomes immediately apparent that the police are treated more favourably than any other emergency service. Discuss.

Answer plan

This question involves a consideration of the liability of the police and the other emergency services, such as the fire brigade, the coastguard service and the ambulance service. It requires a good knowledge of the recent cases in this area. It is also vital that the candidate should not merely list the services and a few decided cases, and state whether or not liability was imposed, but should rather discuss the courts' basis for imposing or refusing to impose a duty of care.

The following points need to be discussed:

- the courts' reluctance to impose a duty of care on the police;
- variability of this rule;
- duty of care and the fire service; and
- duty of care and the coastguard and ambulance services.

Answer

Although there have been a number of cases where the courts have refused to impose a duty of care upon members of the police force, it should be immediately stated that the police do not have any general immunity in tort – see, for example, *Rigby v Chief Constable of Northamptonshire* (1985) and *Swinney v Chief Constable of Northumbria Police* (1996).

The suggestion that the courts are reluctant to impose a duty of care on the police arises in two particular areas, namely, where members of the force are sued in respect of the alleged negligent performance of their functions, and where it is sought to impose liability on the police for the misdeeds of third parties.

In the first of these two situations, the courts have held as a matter of public policy that no duty of care should be imposed upon the police. Thus, in *Calverley v Chief Constable of Merseyside Police* (1989), several police officers alleged that the Chief Constable was vicariously liable for the negligence of the investigating officers who were examining complaints made by members of the public against police officers. The House of Lords held that no duty of care was owed by the investigating officers to the police officers. One of the grounds for the decision was that it was contrary to public policy to impose a duty of care as it might impede full investigations into complaints against the police. Lord Bridge also stated that a duty of care would not be imposed upon a police officer investigating a civilian suspect. The public policy argument was also deployed in *Hughes v National Union of Mineworkers* (1991), where a police officer was injured during an industrial dispute. The officer sued his Chief Constable, alleging that he was negligent in his use and deployment of police manpower and had thus caused the injury. The claim was struck out on the ground that the proceedings disclosed no reasonable cause of action. The High Court held that it was contrary to public policy to impose a duty of care on a senior police officer in a serious public disorder situation.

In *Vellino v Chief Constable of Greater Manchester* (2002), the Court of Appeal held that the police did not owe an arrested person a duty of care to see that he was not injured in a foreseeable attempt to escape police custody, although this decision was based on the *ex turpi causa* principle rather than on public policy.

Note that if a police officer is not acting in an emergency situation, then liability may be imposed for negligent acts (*Knightley v Johns* (1982)). Hence, carelessness in the course of a police officer's duties may in certain circumstances give rise to tortious liability. Also, in *Costello v Chief Constable of Northumbria Police* (1999), it was held that a police officer could be liable for not helping a colleague who was being attacked by a prisoner in a police cell. The Court of Appeal held that the officer owed a duty of care to his colleague, but that it did not follow that such a duty was owed to a member of the public. In a similar vein, in *Waters v Commissioner of Police of the Metropolis* (2000), the

House of Lords, overruling the High Court and the Court of Appeal, refused to strike out an action by a police officer who claimed that the Commissioner owed her duties analogous to those owed by an employer to an employee.

There has been much litigation on the situation where a member of the public has been injured due to the deliberate actions of a third party. One of the leading cases on this second situation is *Hill v Chief Constable of West Yorkshire* (1989). Here the mother of the last victim of a serial killer claimed that the police had not used reasonable care and skill to apprehend the murderer and that, had such care and skill been exercised, the murderer would have been caught before her daughter had been murdered. The House of Lords struck out the claim on the ground of public policy: imposing a duty of care would fetter police discretion. The House stated that while the police owe the public a general duty to enforce the law, that duty was not a duty of care owed to the public to identify and arrest an unknown criminal for that policy reason.

The decision in *Hill* has been applied in a number of cases. Thus, in *Clough v Bussan* (1990), the police were told of a dangerously malfunctioning traffic light, but took no steps to alert motorists. An accident occurred and the plaintiff sued the defendant who issued a third party notice against the police. The third party notice was struck out, following *Hill*, on the grounds that although the police were under a general duty to protect life and property, they did not owe a duty of care to the plaintiff. Similarly, in *Alexandrou v Oxford* (1993), the plaintiff shopkeeper had a burglar alarm which sounded in the local police station when it went off. One evening, the alarm sounded and the police checked the front of the shop, found nothing untoward and left. They carelessly failed to check the rear of the shop where they would have seen signs of forced entry and where the burglars were hiding. When the police left, the burglars naturally also left with a large amount of goods from the shop. The Court of Appeal held that the police did not owe a duty of care to the shopkeeper, on the ground that if it were owed to the shopkeeper it would be owed to all members of the public, and that was contrary to *Hill*. Additionally, it was not in the public interest to impose a duty of care on the police, as it would not result in a higher standard of care, but would significantly direct resources from the suppression of crime. A similar view was voiced in *Hill*.

Also, in *Ancell v McDermot* (1993), it was held that the police owed no duty of care to road users as regards hazards discovered by the police on the road. Again, the diversion of resources argument was invoked.

Perhaps the high point of the judicial reluctance to apply a duty of care to the police where third parties are involved is to be found in *Osman v Ferguson* (1993). In view of the decision of the Court of Appeal and subsequent developments, the facts of this case are important. A schoolteacher formed an unhealthy attachment to a young boy and harassed the boy with false sexual allegations. The teacher changed his surname to that of the boy, and damaged property belonging to the boy's family. He was dismissed from his job, but continued to harass the boy. The police were aware of the situation and the teacher even confessed to a police officer that he feared he would commit a criminally insane act. The teacher deliberately drove into a car in which the boy was a passenger. The police laid an information alleging a road traffic offence, but did not proceed further and did not serve the information. Finally, the teacher shot and severely injured the boy and killed the boy's father. The police were sued in negligence on the grounds that although they had been aware of the teacher's activities for a year, they had failed to interview him or charge him with an offence prior to the shooting. The Court of Appeal held that it was arguable that there was a close degree of proximity amounting to a special relationship existing between the police and the boy's family, but that following *Hill* it was against public policy to impose a duty of care.

A similar decision was reached in *Cowan v Chief Constable of Avon and Somerset Constabulary* (2001), where the police had been called to an incident where a member of the public had been threatened with violence. The Court of Appeal held that the police did not owe a duty of care to prevent an offence being committed against that person. The court held that although there was a sufficiently proximate relationship, it was not just, fair and reasonable to impose a duty of care, as it would be against public policy.

Osman, however, has been considered by the European Court of Human Rights (*Osman v UK* (1998)). The Court held that the blanket immunity absolving the police from negligence actions as regards criminal investigations was contrary to Art 6 of the European Convention on Human Rights. The Court regarded this immunity as being disproportionate to the legitimate aim of maintaining the efficacy of the police, as it did not allow other public policy considerations such as degree of carelessness and gravity of harm suffered to be taken into account.

Thus, in future, especially following the enactment of s 2 of the Human Rights Act 1998, such cases may not automatically hold that no duty of care applies, but may only do so after a consideration of all the facts.

It should be noted that the term 'blanket' immunity is a misnomer. In *Swinney*, an informant gave the police information regarding the killing of a police officer. The person suspected was known by the police to be violent. The informant, not unnaturally, requested total confidentiality regarding her supply of information. Unfortunately, a document containing this information was left in an unattended police car in an area where crime was rife. The car was broken into and the suspect obtained the information, together with the informant's name. Following this, violent threats were made against the informant. The informant sued the police and the Court of Appeal refused to strike out the claim. The court held that a sufficient relationship of proximity existed between the plaintiff and the police, and refused to apply the public policy immunity. Indeed, the court stated that public policy as regards the need to encourage and protect informers was a point against the police in this case.

The police have also failed to obtain immunity in two cases involving suicides. In *Kirkham v Chief Constable of Greater Manchester Police* (1990), the Court of Appeal held that the police were under a duty of care to a man of suicidal tendencies to pass this information on to the prison authorities, so that appropriate precautions could be taken. In *Kirkham*, the deceased was of unsound mind. However, more recently, in *Reeves v Commissioner of Police of the Metropolis* (1998), the House of Lords held that a similar duty of care applied where the deceased was not of unsound mind. In *Orange v Chief Constable of West Yorkshire Police* (2001), this duty was restricted to persons who the police knew or ought to have known presented a suicide threat.

Another recent example of police immunity being refused is *Leach v Chief Constable of Gloucestershire Constabulary* (1999). In this case, the claimant acted as an 'appropriate adult', sitting in on police interviews with mentally disturbed persons. She acted thus in a police interview with a mass murderer. As a result of this gruesome experience, she suffered psychiatric injury. She sued the police on the grounds that the police had asked her to act as an appropriate adult, even though they knew she had no training, and that, unlike the police officers involved, she was not offered any counselling in respect of her traumatic experiences. The Court of Appeal held that the police owed no duty of care to an appropriate adult as regards psychiatric injury suffered, as such a duty might adversely affect the ability of the police to interview suspects effectively. However, the second duty alleged was arguable, as it would not hamper the police's ability to conduct effective interviews if they had to offer counselling to the appropriate adult.

Turning now to the other emergency services, the liability of the fire service was recently considered in three cases: *Capital and Counties plc v Hampshire County Council* (1997); *John Munroe (Acrylics) v London Fire and Civil Defence Authority* (1997); and *Church of Jesus Christ of Latter Day Saints (GB) v West Yorkshire Fire and Civil Defence Authority* (1997). These three cases produced somewhat conflicting decisions at first instance, and were subsequently heard together and reported as *Capital and Counties plc v Hampshire County Council* (1997). The question that concerned the court in each case was whether the fire brigade owe a duty of care to those persons who suffer fire damage.

The Court of Appeal held that the fire brigade were not under a duty of care to either respond to emergency calls or take reasonable care to do so. Thus, if a fire brigade fail to attend a fire or get lost on the way to a fire, they are not liable. The Court of Appeal followed *Alexandrou* here, as they could not distinguish between the police and the fire service in this situation. Thus, so far, the police and the fire service have received identical treatment from the courts.

However, the Court of Appeal went on to consider the liability of the fire service when they did arrive at a fire and attempted to extinguish it. In the *Hampshire* case, it was found as a fact that the fire service had actually made the situation worse by mistakenly turning off a sprinkler system which was helping to extinguish the fire. The Court of Appeal held that a duty of care would be imposed, as the fire service were not just guilty of a failure to act, but had caused positive harm to the plaintiff and this established sufficient proximity to found a duty of care. Similarly, if by a positive careless act the fire brigade substantially increased the risk of a danger which they had not created, they would be liable.

The court went on to reject the claim of all three plaintiffs that, by arriving at the scene of a fire, the fire brigade assumed responsibility to the plaintiffs and thus a duty of care arose. Hence, in the *London* and *West Yorkshire* cases, which did not involve positive careless acts, no duty of care was imposed.

Thus, where the fire service are guilty of omissions only, they are treated by the courts in a similar manner to the police; a positive careless act is required to impose a duty of care on the fire service.

When the court went on to consider whether no duty should be imposed on public policy grounds, we can see some difference in approach between the courts' attitude to the police and the fire service. For an immunity to be granted, it had to be shown that imposing a duty of care would clash with some wider object of the law or interest of the parties, for example, where the possibility of liability would have a negative effect on the performance of the function. Otherwise, the investigation of negligence claims would be open to abuse. The court held that there were no such considerations in the fire service situations. The argument that this would encourage defensive fire fighting and divert resources into litigation was rejected. In rejecting this argument, the court stated that the imposition of liability would not impinge on the mind of a fire service officer in an emergency. This is a surprising statement, as the fear of a negligence action when acting in an emergency situation was exactly the argument used in refusing to impose a duty of care in *Hughes*. Hence, the police in such emergency situations have received more favourable consideration than the fire service.

The liability of several other emergency services have been considered recently by the courts. In *OLL Ltd v Secretary of State for Transport* (1997), the liability of the coastguard service arose. It was alleged that the coastguard owed a duty of care to some canoeists who ran into danger at sea. The High Court held that no duty of care was owed, as the coastguard service was indistinguishable from the fire service, and that they would only be liable if they committed a positive negligent act which caused greater injury to the plaintiff (following *Capital and Counties*).

In *Kent v Griffiths* (2000), the Court of Appeal considered the liability of the ambulance service where an ambulance was called and did not arrive within a reasonable time. The court held that a duty of care did exist. The court distinguished *Capital and Counties plc* on the ground that the ambulance service provided services similar to a hospital, rather than services provided by the police or fire service, whose duty was to protect the general public. Thus, the court reasoned that as doctors can be liable, so should ambulance staff. The court held that the acceptance of the emergency call established the duty of care and placed great weight on the fact that the ambulance was going to collect a known, named person, rather than acting to protect the public generally. Thus, basically, the ambulance service is treated differently from the police, the fire and coastguard services.

Overall therefore it is submitted that, in the past at least, the police have received more favourable treatment than the other emergency services. However, the police do not have a 'blanket' immunity.

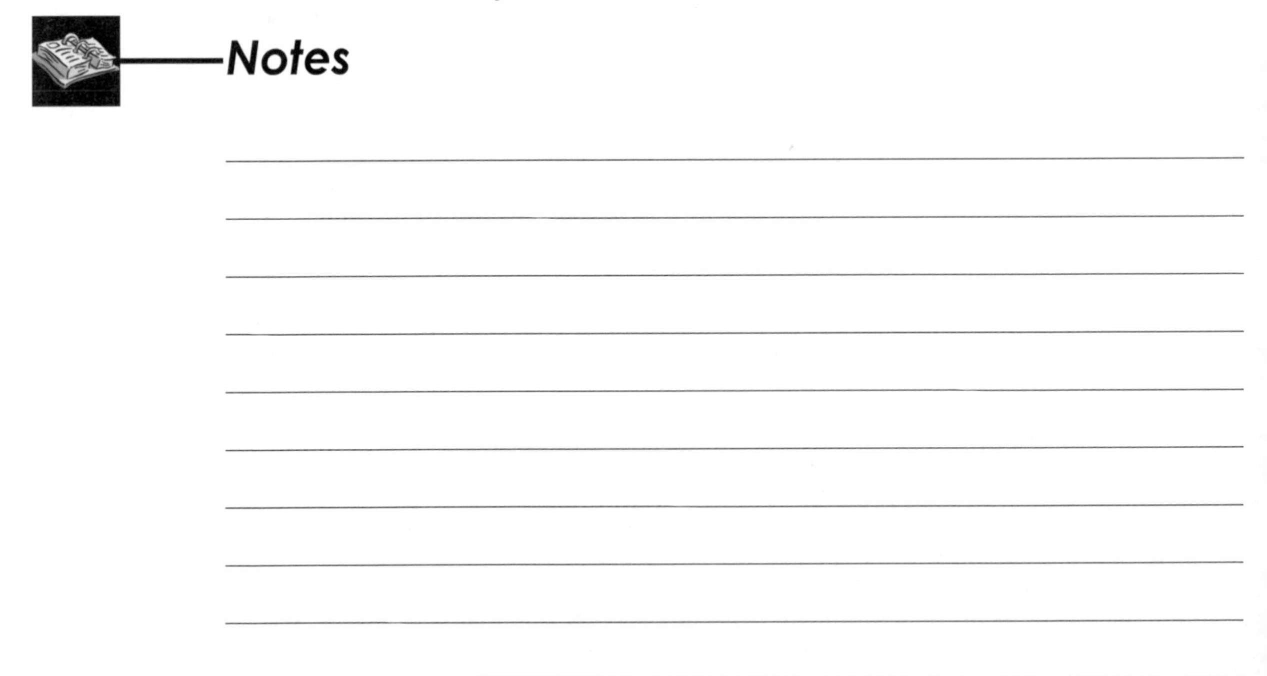

3 Negligence – Breach, Causation and Remoteness of Damage

Introduction

Questions involving breach, causation and remoteness of damage are popular with examiners, either as questions in their own right or as part of a question. Thus, the rule that a tortfeasor takes his victim as he finds him often features as part of a negligence question.

Checklist

Students must be familiar with the following areas:

(a) breach:

- standards and guidelines used to assess whether the defendant's actions are in breach of a duty of care;
- *res ipsa loquitur*;

(b) causation:

- the 'but for' test and the decision in *Fairchild v Glenhaven Funeral Services Ltd* (2002); and

(c) remoteness:

- reasonable foreseeability and the egg shell skull rule;
- *novus actus interveniens*.

Question 11

One day, when walking home, William trips and falls, damaging his knee. Several days later, while driving to work, he sees Victor crossing the road and brakes to avoid running into him. Unfortunately, due to the pain in William's knee, he cannot fully press his brake pedal and as a result he runs into Victor. The collision occurs at a fairly slow speed and a normal person would only have suffered bruising as a result, but Victor has brittle bones and suffers two broken legs and a number of broken ribs. He is taken to the local hospital where, due to an administrative mistake, his right arm is amputated.

Advise Victor.

Answer plan

This is a straightforward question on breach and causation, together with remoteness of damage, the egg shell skull rule and *novus actus interveniens*. As it is relatively simple, care must be taken to discuss the relevant legal principles in depth.

The following points need to be discussed:

- breach of duty by William;
- William takes Victor as he finds him; and
- amputation of the arm – *novus actus interveniens* by the hospital.

Answer

It is well-established law that a road user owes a duty of care to other road users, including pedestrians (*Donoghue v Stevenson* (1932); *Roberts v Ramsbottom* (1980)). Where a duty of care has been previously found to exist, there is no need to apply the modern incremental formulation preferred by the House of Lords in *Caparo Industries plc v Dickman* (1990) or *Murphy v Brentwood District Council* (1990). One could also note the statement of Potts J at first instance in *B v Islington Health Authority* (1991), where he stated that in personal injury cases, the duty of care remains as it was pre-*Caparo*, namely, the foresight of a reasonable person (as in *Donoghue*), a finding that does not appear to have been disturbed on appeal (1992).

As William owes Victor a duty of care, we must consider whether he is in breach of this duty of care. The standard of care required is the objective one of a reasonable person. Thus, in *Blyth v Birmingham Water Works* (1856), Alderson B stated that 'negligence is the omission to do something which a reasonable man, guided upon those conditions which ordinarily regulate the conduct of human affairs, would do, or the doing of something which a prudent and reasonable man would not do'. It is important that the correct question is addressed – the question is not 'did William act reasonably?', but 'what would a reasonable person, placed in his position, have done, and did William meet that standard?'. Applying this objective standard to car drivers, it can be seen that the correct standard to adopt is that of the reasonable, competent driver. Thus, it is irrelevant that a particular driver is a learner (*Nettleship v Weston* (1971)) or, through no fault of his own, he cannot fully control the car for medical reasons and he is otherwise not at fault (*Roberts*). To hold otherwise, as Megaw LJ pointed out in *Nettleship*, would mean adopting a variable standard which could not logically be confined to car drivers and would have to be a universal principle, giving great uncertainty and making it impossible to arrive at consistent decisions. Thus, William must be judged by the standard of the reasonable, competent driver, and he clearly does not meet this standard. The fact that this is due to a medical reason which is outside his control is irrelevant (*Roberts*).

Having decided that William is in breach of his duty, we must determine whether his breach caused Victor's injuries. Turning first to Victor's broken legs and ribs, it is clear, applying the 'but for' test proposed by Lord Denning in *Cork v Kirby MacLean* (1952), that this damage would not have happened but for his breach of duty. Hence William will be liable for Victor's broken legs, provided that the damage is not too remote. The test for remoteness of damage is that the damage must have been reasonably foreseeable (*The Wagon Mound* (1961)). Therefore, the important question is just what damage has to be foreseeable to render that damage not too remote. Also for harm to the person, as long as some personal injury is foreseeable, it does not matter that the exact consequences were unforeseeable (see, for example, *Dulieu v White* (1901); *Smith v Leech Brain* (1962)). Thus, William must take his victim as he finds him, that is, with brittle bones.

We must also consider whether William is responsible for Victor's amputated arm by applying the 'but for' test. As a matter of pure logic, but for William's negligence Victor would not have been at the hospital and the amputation would not have taken place. But we need to consider whether there has been a break in the chain of causation, that is, whether the negligence of the hospital constitutes a *novus actus interveniens*. The new act is that of a third party over which William has no control. To break the chain of causation, it must be something unwarrantable, a new cause which disturbs the sequence of events, and something which can be described as either unreasonable or extraneous or extrinsic (*per* Lord Wright in *The Oropesa* (1943)). Thus, the defendant will remain liable if the act of the third party is not truly independent of the defendant's negligence. It seems in William's case that the act of the hospital does satisfy this criterion. In *Knightley v Johns* (1982), a third party acted negligently. The court held that negligent conduct was more likely to break the chain of causation than non-negligent conduct and that, in *Knightley*, there were so many errors and departures from common sense procedures that the chain of causation had been broken. Looking at the facts of Victor's case, it seems that the hospital has been negligent and there must have been some errors and departures from common sense procedures. Hence, the chain of causation has been broken. Thus William is not liable for the amputated arm; liability for this damage will rest with the hospital.

As Victor was presumably under anaesthetic when his arm was amputated, he may have problems in proving the hospital's lack of care. However, in such a situation, he can rely on the maxim *res ipsa loquitur*, that is, the thing speaks for itself. Where the maxim applies, the court may be prepared to find a breach of duty in the absence of specific evidence of the defendant's actions (see, for example, *Scott v London and St Katherine's Docks* (1865)).

For the maxim to be applicable, it must be shown:

(a) that the defendant is in control of the thing which caused injury to the claimant;

(b) that the accident would not have occurred in the ordinary course of events without negligence; and

(c) that there is no other explanation for the accident.

An example of the maxim in action is *Mahon v Osborne* (1939), where a surgeon left a swab in a patient's body. The application of the maxim will not shift the burden of proof, which will remain on Victor throughout (*Ng Chun Pui v Lee Chuen Tat* (1988)), but it will allow the court to draw an inference of negligence (*Ng Chun Pui*, *per* Lord Griffiths).

Thus, Victor is advised to sue William in respect of his broken legs and ribs, and the hospital in respect of the amputated arm.

Notes

Question 12

Dennis works as a labourer for Hopeless plc and needs to use a ladder to carry out some work. Dennis collects a ladder from Eric in the stores but, when he is halfway up the ladder, steps on a faulty rung and falls to the ground, cutting his shoulder. Dennis goes to his doctor and is given an anti-tetanus injection to which he is allergic. He suffers such an adverse reaction that he is off work for three months without pay. Hopeless plc denies any liability, pointing out that it is a strict company rule that if an employee uses a ladder, he must place a restraining block behind it to ensure that it does not slip, and that Dennis had neglected to do this.

Advise Dennis.

Answer plan

This question covers a variety of topics, including employers' liability, causation and remoteness of damage.

The following points need to be discussed:

- Hopeless' duty to Dennis;
- Hopeless in breach of duty;
- Hopeless' liability for costs and loss of wages;
- liability of Dennis' doctor for loss of wages; and
- the effect of Dennis' non-compliance with the ladder rule.

Answer

It is trite law that Hopeless, as Dennis' employer, owes Dennis a duty of care to take reasonable care for Dennis' safety. In particular, Hopeless owes Dennis a duty to provide properly maintained plant and equipment (*Smith v Baker* (1891)). This is a primary, non-delegable duty that rests with the employer and in addition Hopeless will be vicariously liable for any negligence on the part of Eric while Eric is acting in the course of his employment.

As regards these duties, where a duty of care has been previously found to exist, as in the employer/employee situation (that is, Hopeless plc and Dennis) and between fellow employees (that is, Eric and Dennis), there is no need to apply the modern incremental test to determine the existence of a duty of care (preferred by the House of Lords in *Caparo Industries plc v Dickman* (1990) and *Murphy v Brentwood District Council* (1990)). One could also note the statement of Potts J at first instance in *B v Islington Health Authority* (1991), where he said that in personal injury cases, the duty of care remains as it was pre-*Caparo*, namely the foresight of a reasonable person (as in *Donoghue v Stevenson* (1932)), a finding that does not appear to have been disturbed on appeal (1992).

Taking these two possible causes of action in turn, as the ladder with which Dennis has been supplied is defective, Hopeless is in breach of its duty to provide properly maintained plant and equipment. It would be no defence to Hopeless to allege that it bought the ladder from a reputable supplier and had no reason to suspect that it was defective: s 1 of the Employers' Liability (Defective Equipment) Act 1969. The Act provides that where an employee suffers personal injury in the course of his employment in consequence of a defect in equipment provided by his employer for the purposes of the employer's business and the defect is attributable wholly or partly to the fault of a third party (whether identified or not), the injury shall be deemed to be also attributable to negligence on the part of the employer. In addition, Eric owes Dennis a duty of care under straightforward *Donoghue* principles. It may well be the case that Eric was in breach of his duty by failing to notice that the ladder had a faulty rung. Consequently, as we have the employer/employee relationship between Hopeless and Eric, and Eric is acting in the course of his employment, Hopeless will be vicariously liable for Eric's negligence.

In both situations, as we have shown the existence of a duty of care and a breach of that duty, we need to consider whether the injury suffered by Dennis was caused by the breach. Applying the 'but for' test described by Lord Denning in *Cork v Kirby MacLean* (1952), it seems clear that the injury to Dennis would not have occurred but for the faulty rung, that is, for the breach. The effect of Dennis ignoring the company rule regarding the restraining block is irrelevant to the question of causation, for even if Dennis had complied with this rule, the damage would still have occurred (*Barnett v Chelsea and Kensington Hospital Management Committee* (1968)). In *Barnett*, a man went to the casualty department of a hospital complaining of vomiting. The doctor on casualty duty refused to examine him and sent him home. Some five hours later, the man died from arsenic poisoning. It was held that the doctor was negligent in not examining the man, but that this negligence had not caused the man's death, as even if the doctor had examined and treated him he would still have died because the poisoning could not have been detected and cured in time. Similarly, even if Dennis had placed a restraining block behind the ladder, the faulty rung would have caused his fall to the ground.

We need to see whether all or any of the damage suffered by Dennis is too remote, that is, whether or not the damage is reasonably foreseeable (*The Wagon Mound (No 1)* (1961)). For personal injury, the requirement is that some damage of a foreseeable kind must occur and it is irrelevant that the specific damage suffered cannot be foreseen (*Dulieu v White* (1901); *Smith v Leech Brain* (1962)). It is often said that the tortfeasor takes his victim as he finds him. Clearly, therefore, Hopeless will be liable for the cut to Dennis' shoulder. Next, we must consider whether or not Hopeless is liable for the three months' loss of wages. In *Robinson v Post Office* (1974), the plaintiff was injured at work and suffered an allergic reaction to an anti-tetanus injection. It was held by the court that the defendants were liable for this reaction, because the need for such an injection was reasonably foreseeable and the defendant must take the victim as he finds him. However, it is vital to note that no test for allergic reaction was carried out and, even if it had been done, there would have been no indication of allergy (see *Barnett*). But medical science has advanced from 1974 and if a test is now available that would indicate an allergy in time, Dennis' doctor would be negligent in not carrying out such a test and this negligent act would break the chain of causation.

Where it is alleged that the act of a third party, over whom the claimant has no control, has broken the chain of causation, it must be shown that the act was something unwarrantable, a new cause which disturbs the sequence of events. It must be something which can be described as either unreasonable or extraneous or extrinsic (*per* Lord Wright in *The Oropesa* (1943)). Thus, the defendant will remain liable if the act of the third party is not truly independent of the defendant's negligence. In *Knightley v Johns* (1982), the Court of Appeal held that negligent conduct was more likely to break the chain of causation than non-negligent conduct, and that, in *Knightley*, there were so many errors and departures from common sense procedures that the chain of causation had been broken.

If a test for the allergy exists, then Dennis' doctor is in breach of his duty to Dennis in not carrying out such a test. In *Bolam v Friern Hospital Management Committee* (1957), it was held that in cases of alleged medical negligence, the standard to be applied in determining whether a breach of duty had occurred was that of a reasonably competent medical practitioner. If such a person would have applied an allergy test (if such a test exists) and Dennis' doctor did not, the doctor is in breach of his duty to Dennis. This would amount to such a departure as to break the chain of causation. In this situation, Dennis should sue his doctor in respect of his three months' loss of wages. So long as the doctor's decision was reasonable (or if there was in fact no test), following the authority of *Robinson*, Hopeless is liable for all the harm suffered by Dennis.

One might add that if there were two schools of thought regarding the efficiency or wisdom of carrying out such a test, and Dennis' doctor chose one that held it unwise to administer a test, that would not by itself amount to negligence (*Maynard v West

Midlands Health Authority (1985)). The fact that the decision turned out to be wrong does not prove breach. The question is whether the doctor displayed such a lack of clinical judgment that no doctor, using proper care and skill, could have reached the same decision (*Maynard; Hughes v Waltham Forest Health Authority* (1990)). If an ordinary skilled doctor could have made the same decision, there would be no breach (*Knight v Home Office* (1990)). Although this would seem to allow the medical profession to set its own standards as regards breach, it must still be shown to the satisfaction of the court that the professional opinion Dennis' doctor chose was reasonable or responsible (*Bolitho v City and Hackney Health Authority* (1997)).

Notes

Question 13

> ... two causes may both be necessary preconditions of a particular result ... yet the one may, if the facts justify that conclusion, be treated as the real, substantial, direct or effective cause and the other dismissed ... and ignored for the purposes of legal liability ... [*per* Lord Asquith in *Stapley v Gipsum Mines* (1953)].

Does this statement accurately reflect the law and if so does it allow a judge to choose any previous act as the real cause of the claimant's damage?

Answer plan

This question calls for a discussion of the 'but for' test of causation, and some of the situations in which its application is not straightforward.

The following aspects of causation need to be discussed:

- the 'but for' test;
- pre-existing conditions;
- successive causes; and
- *novus actus interveniens*.

Answer

The test that the courts usually use in deciding whether or not a particular act was the cause of the claimant's injury is the 'but for' test. The test was elucidated by Lord Denning in *Cork v Kirby MacLean* (1952), where he said that 'if the damage would not have happened but for a particular fault, then that fault is the cause of the damage; if it would have happened just the same, fault or no fault, the fault is not the cause of the damage'.

A good example of this test is provided by *Barnett v Chelsea and Kensington Hospital Management Committee* (1969), where a man went to the casualty department of a hospital complaining of vomiting. The doctor refused to examine him and sent him home. Some five hours later, he died from arsenic poisoning. It was held that the doctor was negligent in not examining the man, but that his negligence had not caused the man's death, as even if the doctor had examined and treated him he still would have died because the poisoning could not have been detected and cured in time.

Although the 'but for' test works well in the vast majority of cases, it does give rise to problems in some situations, especially where there is more than one possible cause of the claimant's loss. Thus, where the claimant's loss is due to a pre-existing condition, rather than to the defendant's actions, the defendant may only be liable for part of the damage suffered by the claimant. In *Cutler v Vauxhall Motors* (1971), the plaintiff suffered a graze to his ankle due to the negligence of the defendants. The plaintiff had an existing varicose vein condition and as a result of the graze it was decided to operate immediately to cure this condition. It was held that the plaintiff could recover for the graze, but not for the operation, as the varicose vein condition would have required an operation at some time in the future in any event. *Performance Cars v Abraham* (1962) is an example of a pre-existing condition working in favour of the plaintiff, rather than against him as in *Cutler*. In both *Cutler* and *Performance Cars*, the pre-existing condition was treated as the effective cause of part of the plaintiff's loss.

Another area in which the 'but for' test can give rise to problems is where there is more than one cause of the claimant's injury, for example, where two persons both cause harm to the claimant, so that he still would have suffered harm but for the negligence of either of the defendants. In such a situation, the 'but for' test would mean that neither defendant was liable to the claimant, but a court would not reach such a conclusion in practice. This situation was recently considered by the Court of Appeal in *Holtby v Brigham and Cowan (Hull) Ltd* (2000). Here, the claimant suffered injury as a result of exposure to a noxious substance by two or more persons, but claimed against one only. The Court of Appeal held that the defendant was liable, but only to the extent that he had caused the claimant's injury. The courts tend to be rather proud of the fact that they approach causation as a matter of common sense, rather than from any academic or theoretical point of view. As Lord Wright stated in *Yorkshire Dale Steamship v Minister of War Transport* (1942), 'causation is to be understood as the man in the street, and not as either the scientist or the metaphysician would understand it'.

This common sense approach to causation can be seen in those situations in which another act has occurred after the original negligent act of the defendant, that is, the *novus actus interveniens* situation. The judge must then decide which of the two acts is the real, substantial, direct or effective cause. The *novus actus* may be either an act of the claimant or of a third party or of nature. Taking these in turn, the latter act of the claimant which causes additional harm may be held to have broken the chain of causation between the original negligent act of the defendant and the additional harm suffered by the claimant. For example, this latter act of the claimant may be treated as the real or effective cause of the claimant's additional loss and the original negligent act of the defendant ignored for the purposes of the additional liability. However, a judge does not have a completely free choice in deciding whether or not this latter act is the effective cause of the additional harm; the decided cases lay down a rule that the latter act of the claimant will only be held to be the true cause if the additional harm is caused by an act which is unreasonable. An example is *McKew v Holland and Hannen and Cubitts* (1969), where the plaintiff, as a result of the defendants' negligence, occasionally lost control of his leg. Despite this injury, the plaintiff still went down a steep flight of stairs which had no handrail and fell when his leg gave way. The House of Lords held that he could not recover for this injury. The House held that this act was so unreasonable that the original negligence of the defendants could be ignored for the purposes of legal liability. In contrast, consider *Wieland v Cyril Lord Carpets* (1969), where the plaintiff, as a result of the defendants' negligence, was unable to use her bi-focal spectacles in the normal manner. As a result of this she fell down a flight of stairs. It was held that the defendants were liable for this additional harm to the plaintiff because she had not acted unreasonably in continuing to wear her bi-focals. Thus, the court has a guideline in deciding whether to allow recovery for the latter damage suffered by a claimant. However, as the guideline involves a decision as to the reasonableness or otherwise of a claimant's (or other party's) conduct, it will often give the judge a certain amount of discretion.

Where the latter act is that of a third party, this latter act will be treated as the real cause of the claimant's additional damage where it is something 'ultroneous, something unwarrantable, a new cause which disturbs the sequence of events, something which can be described as either unreasonable or extraneous or extrinsic' (*per* Lord Wright in *The Oropesa* (1943)). Thus, the latter act of the third party will not be treated as the true cause of the additional damage unless it is independent of the defendant's original negligence. If the act of the third party is itself negligent, the courts are usually willing to hold that this act is the true cause of the claimant's additional damage (*Knightley v Johns* (1982)). In the case of *Wright v Lodge* (1993), the Court of Appeal held that a driver who is involved in a collision partly due to his own negligence could be exonerated for responsibility for subsequent events which occurred because another driver drove recklessly if those events would not have occurred had that other reckless driver merely been negligent. Again, a guideline is available to a judge, but the decision as to whether the actual latter act is unreasonable or independent will involve a certain amount of discretion. It should perhaps be noted here that a defendant cannot rely on his own additional wrong action to break the chain of causation: *Normans Bay Ltd v Condert Brothers* (2004).

Finally, the latter act may be an act of nature, such as a violent storm, and if it is independent of the original negligence of the defendant, the defendant will not be liable for the additional consequences (*Carslogie Steamship v Royal Norwegian Government* (1952)). Again, a test is available to the judge and it will also involve a certain amount of discretion.

An additional restriction on a judge's freedom to choose which previous action was the real cause of the claimant's injury is the need for the claimant to prove causation. A claimant who has difficulties in this area will usually rely on the decision of the House of Lords in *McGhee v National Coal Board* (1973). This case is authority for the proposition that a claimant may recover if he can show that the actions of the defendant materially increased the risk of damage occurring. The House of Lords took a restrictive approach to *McGhee* in the later cases of *Kay v Ayrshire and Arran Health Board* (1987); *Hotson v East Berkshire Area Health Authority* (1987); and *Wilsher v Essex Area Health Authority* (1988). Indeed, in *Wilsher*, Lord Bridge stated that *McGhee* 'laid down no new principle of law whatsoever. On the contrary, it affirmed the principle that the onus of proving causation lies on the plaintiff'. However, in *Holtby v Brigham and Cowan Ltd* (2000), the Court of Appeal applied the traditional *ratio* of *McGhee* and in *Fairchild v Glenhaven Funeral Services* (2002), the House of Lords emphatically reinstated *McGhee* and held that the statement of Lord Bridge did not accurately reflect the decision in *McGhee* and should no longer be treated as authoritative.

Thus, a judge does have a certain amount of discretion in approaching causation, and indeed in *Fairchild,* the House of Lords recognised that in applying *McGhee* rather than requiring strict proof of causation they were making a policy decision to arrive at a fair result for the claimant. (A similar policy decision regarding causation was reached by the House of Lords in *Chester v Afshar* (2004).)

4 Breach of Statutory Duty

Introduction

Questions on breach of statutory duty often appear in examinations and usually involve a consideration as to whether a breach of statutory duty gives rise to a cause of action in tort. Such questions can also contain issues such as causation, together with employer's liability and contributory negligence.

Checklist

Students must be familiar with the following areas:

(a) whether breach gives rise to a tort:

- presumption that enforcement provided by statute is exclusive;
- exception to presumption where the statute is enacted for the benefit of a class and where the claimant has suffered harm in excess of that suffered by the public at large; and

(b) if breach does give rise to a tort, note that the claimant must prove that:

- the action caused harm of a type regulated by statute;
- the claimant is a person the statute was intended to protect;
- the harm suffered is of the kind the statute was intended to protect.

Question 14

Regulations made under the (fictitious) Oil Products (Protection of Workers) Act 1987 provide, *inter alia*, that 'Employers shall ensure that all workers engaged in the manufacture of oil products wear the protective clothing prescribed in these Regulations when they are at work or likely to come into contact with oil products, and shall ensure that such clothing is maintained in a good state of repair'. These Regulations apply to the premises of Refiners plc.

One day, Alan, who works directly with oil products, puts on a pair of protective overalls but, because he cannot be bothered to take his safety boots off, he rips them down the leg. He replaces these torn overalls on a hanger and puts on a fresh pair. Shortly afterwards, Brian, who works in the accounts section, goes into an area where it is necessary to wear overalls. He puts on the torn overalls without noticing the defect and, whilst in the oil product area, he trips over the torn leg of the overalls and falls, injuring his elbow.

Advise Brian.

Answer plan

This is a typical breach of statutory duty question and the following points need to be discussed:

- whether the breach gives rise to a tort;
- whether the harm is of a type intended to be prevented by statute;
- employer's liability of Refiners; and
- Alan's liability and Refiners' vicarious liability.

Answer

Brian has three possible causes of action against Refiners. The first is for breach of statutory duty, the second is for breach of their duty as employers and, thirdly, Refiners may be vicariously liable for Alan's negligence. We shall consider each possible action in turn.

As regards possible liability for breach of statutory duty, the Regulations provide that the protective clothing worn by Brian 'must be maintained in a good state of repair'. The important question is just what standard of care the Regulations impose on Refiners. Typically, statutory obligations are either subject to a phrase such as 'so far as is reasonably practicable', when they usually add little to the common law of negligence, or are absolute, when the only question to be decided is whether the statutory regulations have been complied with or not. The reasons for not complying will be irrelevant to liability in the latter case. An example of absolute liability is the duty to fence dangerous machines imposed by s 14 of the Factories Act 1961 (*John Summers v Frost* (1955)). In the instant case, the requirement is that the employer shall ensure that such clothing is maintained in a good state of repair and, as there is no mention of reasonableness in the Regulations, the obligation is absolute and so Refiners is in breach of its statutory duty. The next question to be decided is whether or not this breach gives rise to an action in tort. The correct test is to see whether the Regulations on their true construction confer upon Brian a right of action in tort (*Cutler v Wandsworth Stadium* (1949); *X v Bedfordshire County Council* (1995)). If the Regulations address this point as, for example, s 5(1) of the Guard Dogs Act 1975, which expressly excludes civil actions for breach of this Act, that disposes of the matter. Alternatively if, as is usual, the statute (or Regulations) is silent on the point, the court must ascertain the intention of Parliament. In the House of Lords in *Lonrho v Shell Petroleum* (1982), Lord Diplock stated that the initial presumption was that where the statute created an obligation, together with a means of enforcing that obligation (for example, by a criminal penalty), the obligation cannot be enforced in any other way. We are not told of any such means in the Regulations, but even so, the presence of a criminal penalty would not necessarily be fatal to Brian's case. In *Atkinson v Newcastle Waterworks* (1877), the imposition of a fine for breach of a statutory duty was held to be exclusive, whereas in *Groves v Lord Wimborne* (1898), the provision in the Regulations of a fine for breach was held not to deny the plaintiff a cause of action. It is worth noting in this respect that in *Groves*, as in the present case, the statute was enacted for the benefit of a class. The absence of any such provision would make it easier for Brian to claim that a right in tort existed (*Thornton v Kirklees Metropolitan Borough Council* (1979)). Lord Diplock continued to state that there were two exceptions to this general rule and one is relevant here, namely, when the statute is enacted for the benefit of a class of persons and the claimant is a member of that class. As employees are regarded as a class of persons for whose benefit industrial safety legislation is enacted, as in *Groves*, it would seem that *prima facie* Brian can sue in respect of Refiners' breach of statutory duty.

However, Brian has only cleared the first hurdle here. The next matter he must prove is that the act which caused the harm is regulated by statute, that he was one of the persons the statute was intended to protect, and that the harm suffered was of a kind that the statute was intended to prevent.

Brian should have no problem with the first two requirements, but he will have a problem with the third. It seems from the Regulations that the requirement to provide protective clothing in good condition was to stop oil products from coming into contact with a person's body and not to prevent tripping or falling. In *Gorris v Scott* (1874), a shipowner was required by statute to provide pens on board his ship for cattle. He failed to do this and the plaintiff's cattle were swept overboard. It was held that the shipowner was not liable, because the harm the statute was intended to prevent was the spread of contagious diseases, and not to prevent the cattle being swept overboard. Similarly, in *Fytche v Wincanton Logistics plc* (2004), the House of Lords held that an employer who issued an employee with boots with steel toecaps for protection from impact injuries was not liable for an injury caused by a defect in the boots that had no effect on impact protection.

By analogy with *Gorris* and *Fytche*, it would seem that Brian cannot bring himself under the ambit of the statute.

We must next consider whether Brian can sue Refiners in negligence. As Brian's employer, Refiners owes Brian a duty of care to provide proper plant and equipment (*Smith v Baker* (1891)). Clothing comes under the definition of equipment (see, for example, s 1(1) of the Employers' Liability (Defective Equipment) Act 1969). However, this duty is not an absolute one, but merely one to take reasonable care for the employees' safety. In *Toronto Power v Paskwan* (1915), Sir Arthur Channell stated that 'if in the course of working plant becomes defective and the defect is not brought to the master's knowledge and could not by reasonable diligence have been discovered by him, the master is not liable'. As we are told that Brian put on the overalls 'shortly afterwards', it would seem that Refiners is not in breach of its duty regarding equipment. Brian could perhaps attempt to show that Refiners is in breach of its duty to provide a safe system of work, in that it has failed to provide a disposal system for torn overalls and a sufficient quantity of overalls in good condition. Refiners could reply that it does normally meet these two requirements of a safe system of work and that it was the action of Alan in replacing the overalls that was the cause of the harm. However, in *McDermid v Nash Dredging and Reclamation* (1987), the House of Lords held that the duty of the employer was not just to provide a safe system of work, but to ensure that a safe system was actually operated. Such a duty is non-delegable, and it would be no defence for Refiners to show that it delegated performance to an employee who it reasonably believed to be competent to perform it (*per* Lord Brandon).

It is clear that Alan himself owes Brian a duty of care under normal *Donoghue v Stevenson* (1932) principles, in that he can reasonably foresee that any lack of care on his part may cause injury to Brian. There is no need to apply the modern incremental formulation preferred by the House of Lords in *Caparo Industries plc v Dickman* (1990) and *Murphy v Brentwood District Council* (1990). Indeed, in *B v Islington Health Authority* (1991), at first instance, Potts J stated that in personal injury cases, the duty of care remained as it was pre-*Caparo*, namely, the foresight of a reasonable person (as in *Donoghue*), a finding that does not appear to have been disturbed on appeal (1992). Alan will be in breach of this duty if a reasonable person, placed in his position, would not have acted in this way (*Blyth v Birmingham Waterworks* (1856)). It is submitted that a reasonable person would not have replaced the torn overalls on the hanger, but would have disposed of them in a safe manner. This breach must have caused Brian's injury, and the 'but for' test in *Cork v Kirby MacLean* (1952) proves the required causal connection. Additionally, the injury suffered by Brian must not be too remote, in that it must be reasonably foreseeable (*The Wagon Mound (No 1)* (1961)). All that Alan need foresee is some personal injury; he need foresee neither the extent (*Smith v Leech Brain* (1962)) nor the exact manner in which the damage occurs (*Hughes v Lord Advocate* (1963)). All these criteria are satisfied, and so Alan has been negligent as regards his conduct to Brian. As Alan is an employee of Refiners, and was acting in the course of his employment when this negligence took place, it follows that Refiners is vicariously liable for Alan's negligence.

It could be argued that neither Refiners' breach of statutory duty nor Alan's breach of common law duty caused Brian's injury; rather it was Brian's carelessness in failing to note the damaged overall that caused the injury, that is, that this action by Brian constituted a *novus actus interveniens* which broke the chain of causation. A subsequent act of the claimant may amount to a *novus actus* where his conduct has been so careless that his injury can no longer be attributed to the negligence of the defendant. An examination of the two leading cases in this area, *McKew v Holland and Hannen and Cubitts* (1969) and *Wieland v Cyril Lord Carpets* (1969), shows that the test the courts apply is whether the claimant's conduct was reasonable or not and, if it is unreasonable, it will break the chain of causation. It does not seem unreasonable of Brian, who normally has no need to wear overalls, to assume that those provided by Refiners are in good condition. Also, as regards Refiners' breach of statutory duty, in *Westwood v Post Office* (1974), it was held that the fact that the plaintiff was himself at fault did not allow the defendants to act in breach of their statutory duty, and that the plaintiff was entitled to assume that the defendants would comply with their statutory obligations. Thus, *Westwood* would dispose of this argument (if it were to be decided that this case fell outside of *Gorris*). It is more likely that a defence of contributory negligence might succeed in reducing Brian's damages if it could be shown that Brian had taken insufficient care for his own safety (*Jones v Livox Quarries* (1952)). As there seems to be no emergency as in *Jones v Boyce* (1816), contributory negligence cannot be ruled out.

Notes

Question 15

'Until recently the chances of successfully suing a local authority for breach of statutory duty appeared to be slim. However, recent developments in this area of law, and particularly the effect of the Human Rights Act 1998, have substantially increased the chances of a plaintiff succeeding.'

Discuss the above statement.

Answer plan

This question requires an essay on breach of statutory duty on the part of local authorities. This is a highly complex area of law, and such a question should be attempted only by candidates who have a good, up to date grasp of case law post-*X v Bedfordshire County Council* (1995).

The following points need to be discussed:

- ingredients of the tort of breach of statutory duty;
- application of these principles in *X v Bedfordshire County Council* (1995);
- judicial retreat from *X* and the effect of the Human Rights Act.

Answer

A breach by a defendant local authority of a duty imposed on the authority by statute may give rise to a cause of action in tort. The main problem which arises in this area is to decide which statute, when breached, gives rise to such an action. In some cases, it may be clear that an action will arise, as that is one of the purposes of the statute. In other cases, the statute may expressly exclude an action in tort upon breach, for example, s 5 of the Guard Dogs Act 1975. Unfortunately, in the great majority of cases, the statute says nothing as to whether or not a breach will give rise to an action in tort. The courts have therefore developed a set of principles to answer this question.

The general approach adopted by the courts is to decide whether the statute, on its true construction, gives the claimant the right to sue in tort for a breach of the statute (*Cutler v Wandsworth Stadium Ltd* (1948)). Thus, the function of the courts here is to discern the intention of Parliament. In *Lonrho v Shell Petroleum* (1982), Lord Diplock stated that the initial presumption was that where the statute created an obligation together with a means of enforcing that obligation (for example, a criminal penalty), the obligation cannot be enforced in any other way. Lord Diplock went on to state that there were two main exceptions to this basic presumption: first, where the statute was enacted for the benefit of a particular class of persons; and secondly, where the statute created a public right and the claimant suffered harm over and above the harm suffered by the public at large. In *CBS Songs v Amstrad Consumer Electronics plc* (1988), the Court of Appeal held that this classification was comprehensive, and that no other exceptions existed to the basic presumption. In cases involving local authority defendants, the relevant statutes are usually treated as having been passed for the benefit of the public generally, and not for a particular class of persons.

The whole area of breach of statutory duty was reviewed by the House of Lords in *X v Bedfordshire County Council* (1995), which consisted of five consolidated appeals based on statutes which were enacted to protect children from abuse and statutes which were enacted for children with special educational needs. The defendant local authorities sought to strike out the claims as disclosing no cause of action. The House of Lords held that the abuse claims and the education claims had been correctly struck out, as the relevant statutes showed no clear evidence of the intention of Parliament to allow an action for breach of statutory duty. However, the House of Lords also had to consider claims arising from the same set of facts in common law negligence, and the negligence actions were struck out in the abuse claims but not in the education claims. Although such a negligence claim differs from a breach of statutory duty claim, we need to consider the negligence claims, as they represent another route through which a defendant local authority may be held liable on the same facts as those arising in a breach of statutory duty action.

In considering breach of statutory duty, Lord Browne-Wilkinson, who gave the leading judgment in *X*, accepted that in the abuse cases, the legislation in *X* was passed for the protection of a limited class, namely, children at risk. These are clearly factors which point to the existence of a right to sue for breach of the statute. However, it was held that these factors were outweighed by other matters. First, that the legislation established a system to promote the social welfare of the community. Secondly, many of the duties imposed were expressed in broad language, with much being left to the subjective judgment of the local authority. Finally, when considering the sections which gave rise to the alleged breach, the House found it impossible to construe these sections as showing an intention of Parliament that the authority should be liable in damages if the court decided with hindsight that there had been a breach by the local authority. In one of the educational cases, the House was prepared to find that a child with special educational needs was a member of a class for whose protection the statute was enacted. However, the House thought that the language of the statute was far too general and vague to give rise to an action for breach of statutory duty. In addition, the statute provided in great detail for a procedure relating to children with special needs; it provided for the close involvement of all those persons affected by any decision and gave extensive rights of appeal. Taking the legislation as a whole, the House found that it was impossible to find that Parliament had intended to give the right to sue for damages as well.

An alternative approach for a potential claimant is to consider whether liability could arise on the part of the local authority for negligent exercise of a statutory power or discretion. The courts traditionally have been reluctant to impose such liability, and in *Home Office v Dorset Yacht Co* (1970), the House of Lords held that liability would arise only where the authority's exercise of its statutory discretion was *ultra vires* the authority. In the *Bedfordshire* case, the House stated that the *ultra vires* doctrine was irrelevant to these situations, and that the court should decide whether the decision lay outside the ambit of the discretion altogether. If the decision did lie outside the discretion, it was possible, but not necessarily automatic, that a common law duty of care would be imposed. As an additional factor, some matters were just inappropriate for a court to decide. In this context, the House of Lords followed the distinction between 'policy' and 'operational' decisions drawn in *Anns v Merton London Borough Council* (1977). It stated that if the decision involved a question of 'policy', it was not appropriate for the court to decide it. Thus, liability would not be imposed upon local authorities unless the discretion fell within the 'operational' as opposed to the 'policy' area and if the decision fell outside the discretion given by the statute.

In the actual *X* case, the House refused to strike out the actions on the above grounds, as in neither the abuse nor the educational cases did the allegations of negligence require the court to consider policy matters.

Once the decision that the cases could be considered by the court had been reached, the House had to decide whether a duty of care arose under the principles of *Caparo Industries plc v Dickman* (1990), that is, whether in all the circumstances it was fair, just and reasonable to impose a duty of care. The House started with the basic premise that wrongs should be remedied, and that very powerful considerations are required to override that policy. In the instant cases, it was held that such considerations did exist. In the abuse cases, there was a pre-existing statutory system in place; the imposition of a duty of care would make the local authority unduly cautious in the exercise of its powers; and, finally, the existence of alternative remedies supported the case against imposing a duty of care. In the educational cases, no duty of care was imposed on the local authority for similar reasons.

As well as direct actions against the local authority, indirect actions were brought alleging that the authorities were vicariously liable for the actions of their employees. In the abuse cases, the House of Lords struck out these actions, but refused to strike them out in the educational cases. Their Lordships' rationale for this distinction was that in the abuse cases, the employees were employed (in part) to care for the plaintiffs.

The effect of the *X* case was to make it difficult for claimants to succeed against local authorities in a direct action, either for breach of statutory duty or in negligence. This restrictive approach was continued in *Stovin v Wise* (1996). In this case, the claim was based on the local authority's failure to exercise a statutory power. The House of Lords held that in such a case, the fact that Parliament had chosen to confer a discretion rather than to impose a duty indicated that the policy of the relevant Act was not to give a right to compensation.

Thus, following this line of authority, in *H v Norfolk County Council* (1997), it was held that a local authority was not liable for failing to take reasonable steps to prevent a child in care being abused by foster parents. Also, in *Barrett v Enfield London Borough Council* (1997), the Court of Appeal held that a local authority owed no duty of care to a child who suffered psychiatric illness after being moved nine times between different foster homes. However, when *Barrett* (1999) was heard in the House of Lords, the possibility of a wider liability of local authorities emerged from that decision. In *Barrett*, the House stated that the courts had to keep within reasonable bounds claims against local authorities exercising statutory powers in the social welfare field. However, the House stressed the importance of setting reasonable bounds to such immunity. In deciding whether a particular issue can be decided by the courts, the two guides are discretion and the policy/operational divide. However, the House emphasised the fact that merely because some discretion is involved in an act under a statutory power, that does not automatically mean that common law negligence is ruled out. Thus, *Barrett* in the House of Lords is clear authority that only where the decision would involve the court in having to consider matters of policy will the claimant's claim be struck out. If such matters are absent, the court should decide the matter under common law negligence. If the court does decide that a duty of care is owed, the manner in which the discretion was exercised will be relevant in deciding whether there was a breach of duty.

In *Barrett*, the House of Lords also considered the just, fair and reasonable criteria for imposing a duty on local authorities. In *X*, the House of Lords had given a number of reasons for finding that the criteria were not satisfied. In *Barrett*, the House of Lords considered these reasons and rejected them. There seem to be no valid reasons for preferring the view of the House in *X* to the view of the House in *Barrett*. In both cases, a series of statements were made with no evidence in support. However, it seems very likely that *Barrett* represents a shift away from blanket immunity and thus in favour of claimants. The reason for this movement has undoubtedly been the decision of the European Court of Human Rights in *Osman v UK* (1998). In *Osman*, the Court held that the claimant's right to a fair hearing under Art 6 of the European Convention on Human Rights had been breached, because the English Court of Appeal had struck out Mr Osman's action against the police on the basis of a blanket immunity. Since the coming into effect of the Human Rights Act 1998, this argument is now overwhelmingly powerful. Thus, it seems unlikely that actions against a local authority will be automatically struck out, but will proceed to full trial. Even if a duty of care can be established at trial, the claimant will still have to overcome the problem of proving both breach and causation, so the movement in favour of the claimant in these situations may have practical limitations.

In four appeals heard post-*Barrett* by the House of Lords, reported together in *Phelps v Hillingdon London Borough Council* (2000), the House refused to strike out the cases. The House held that educational psychologists, education officers and teachers could owe a duty of care to a specific pupil (though they did not decide whether on the facts such a duty existed). Having decided this, their Lordships went on to consider whether there were any reasons why the local authority should not be vicariously liable for any breach of such duties. Following *Barrett*, the House decided against a blanket immunity. Again, in *Phelps*, a series of assertions were made with no accompanying evidence, and in this case the assertions were closer in spirit to *Barrett* than to *X*.

In *D v East Berkshire Community Health Trust* (2003), the Court of Appeal reviewed the decision of the House of Lords in *X*, and held that this decision could not stand following the Human Rights Act 1998. In *D*, which consisted of three appeals, parents claimed damages for psychiatric damage caused by false accusations of child abuse. The allegations and subsequent injury all occurred before the Human Rights Act 1998 came into force.

The Court of Appeal held that following *Barrett* and *Phelps*, *X* was only authority for the proposition that local authority decisions as to whether or not to take a child into care were not reviewable via a claim in negligence as it was not fair, just and reasonable to impose a duty of care. This reduced the *ratio decidendi* clearly applied to the appeals in *D*, and so the court had to consider the effect of the Human Rights Act 1998.

The claimants alleged that in striking out the claims, their right to access to the courts under Art 6 of the European Convention on Human Rights was denied, following the arguments in *Osman*. However, in *Z v UK* (2001), where the claimants in *X* appealed to the European Court, the European Court held that as the House of Lords in *X* had decided as a matter of law that no duty of care existed, Art 6, which guaranteed only procedural rights, did not apply. Given that the House of Lords had recently applied *Z* in *Matthews v Ministry of Defence* (2003), the decision in *Z* was fatal to the claimants.

The claimants then claimed that the striking out decision interfered with their right to respect for family life under Art 8 of the European Convention on Human Rights. Unfortunately for the claimants all the relevant events had occurred before the coming into force of the Human Rights Act. Had the events occurred when the 1998 Act was in force the claimants could have relied on *TP and KM v UK* (2001), in which the European Court held that an incorrect decision of a local authority to remove a child from her mother constituted a breach of Art 8. Thus, for events occurring after 2 October 2000 (when the 1998 Act came into force), future claimants could rely on *TP and KM* and Art 8, and could sue local authorities under s 6(1) of the 1998 Act, which makes it unlawful for a public authority to act in contravention to a Convention right.

Given the above conclusion, the Court of Appeal in *D* held that the policy reasons stated by the House of Lords in *X* to decide that no duty of care in negligence was owed to a child now ceased to apply, and that children could bring claims against local authorities regarding decisions as to whether a child should be taken into care.

The decision in *D* applied only to children. It would not be sensible to impose upon a local authority a duty of care in respect of a child and at the same time impose a duty of care in respect of a parent where that parent had been accused of abusing the child. Clearly the interests of the child and parent may be in conflict and so the court held there were valid public policy reasons for not imposing a duty of care as regards parents in these child care situations.

Notes

5 Employers' Liability

Introduction

Questions on employers' liability are often set in examinations. As the topic is only a specialised branch of the law of negligence, it does not introduce any new legal concepts, but generally tests such areas as breach, causation, remoteness, contributory negligence and vicarious liability. The topic may be combined with breach of statutory duty. Students should refer to Chapters 1, 3 and 4 for examples of questions that involve an element of employers' liability. The non-delegable nature of employers' duties should be noted, especially where independent contractors are involved.

Checklist

Students must be familiar with all of the above topics, and especially:

(a) provision of competent fellow employees;

(b) provision of safe plant and equipment;

(c) provision of safe place of work; and

(d) provision of safe system of work.

Question 16

Ken is employed by Lomad plc as an electrician. One day, he is asked to repair a ceiling fan located in Lomad's workplace and is told to dismantle the fan and take it to the electrical workshop for repair. In order to save time, Ken attempts to repair the fan whilst standing on a stepladder and whilst doing so he drops a pair of pliers, which lands on Martin's head. Because Martin is of a rather nervous disposition, he is off work for two months following this accident, rather than the two days which would be normal for such an injury. Following this incident, Ken decides to comply with his instructions and dismantles the fan but, while he is doing this, his screwdriver snaps and a piece of metal enters his eye.

Advise Martin and Ken of any remedies available to them.

Answer plan

This is a straightforward question on employers' liability, involving issues of both primary and secondary liability on the part of the employer.

The following points need to be discussed:

- vicarious liability for Ken's action;
- the egg shell skull rule; and
- the duty to provide a safe place of work.

Answer

Martin will wish to sue Lomad for the harm that he has suffered, and he can sue it in respect of its primary liability to him as his employer, and its secondary liability as being vicariously liable for the negligence of Ken. As regards Lomad's primary liability, Lomad has a duty to provide Martin with a safe place of work. (This is not an absolute duty, but merely places on the employer the duty to take reasonable steps to provide a safe place of work (*Latimer v AEC* (1953); see also *Gitsham v Pearce* (1991) for a more recent example).) We must decide therefore whether Lomad has taken such reasonable steps. It has of course instructed Ken to take the fan to the electrical workshop to repair it, but the problem for Lomad is that the duty to provide a safe place of work is non-delegable. In other words, the employer may entrust the performance of this work to an employee, but he cannot thereby discharge his duty. In *McDermid v Nash Dredging and Reclamation* (1987), Lord Brandon said: 'The essential characteristic of the (non-delegable) duty is that, if it is not performed, it is no defence for the employer to show that he delegated its performance to a person, whether his servant or not his servant, whom he reasonably believed to be competent to perform it. Despite such delegation, the employer is liable for the non-performance of the duty.' Thus, following *McDermid*, we can see that Lomad is in breach of its duty to provide a safe place of work.

Considering Lomad's secondary liability, as Ken is employed by Lomad, Lomad will be liable for any tort committed by Ken in the course of his employment. It is clear that Ken himself owes Martin a duty of care under normal *Donoghue v Stevenson* (1932) principles, in that he can reasonably foresee that any lack of care on his part may cause injury to Martin. There is no need to apply the modern incremental formulation preferred by the House of Lords in *Caparo Industries plc v Dickman* (1990) and *Murphy v Brentwood District Council* (1990). Indeed, in *B v Islington Health Authority* (1991), at first instance, Potts J stated that in personal injury cases the duty of care remained as it was pre-*Caparo*, namely, the foresight of a reasonable person (as in *Donoghue*), a finding that does not appear to have been disturbed on appeal (1992). Ken will be in breach of this duty if a reasonable person placed in his position would not have acted in this way (*Blyth v Birmingham Waterworks* (1856)), and it is submitted that a reasonable person would not have dropped a pair of pliers. This breach must have caused Martin's injury, and the 'but for' test in *Cork v Kirby MacLean* (1952) proves the required causal connection. Additionally, the harm suffered by Martin must not be too remote, in that it must be reasonably foreseeable (*The Wagon Mound (No 1)* (1961)). All that Ken need foresee is some personal injury; he need foresee neither the extent (*Smith v Leech Brain* (1962)) nor the exact manner in which the damage occurs (*Hughes v Lord Advocate* (1963)). All these criteria are satisfied, and so Ken has been negligent as regards his conduct to Martin.

As Ken is an employee of Lomad and was acting in the course of his employment when this negligence took place, it follows that Lomad is vicariously liable for Ken's negligence. We are told that Ken is an electrician, and in repairing the fan he is *prima facie* acting within the course of his employment. However, we need to consider the effect of the express prohibition that he should not repair the fan *in situ* and whether, by acting in contravention of this prohibition, he has stepped outside the course of his employment. The authorities show that acting in contravention of a prohibition will not automatically take the act outside the course of the employment, for example, *Rose v Plenty* (1976) and *Limpus v London General Omnibus* (1862). What a prohibition can do is to limit those acts which lie within the course of the employment, but it cannot restrict the mode of carrying out an act that does lie within the course of the employment (see, for example, *Limpus*). Thus, the question that must be decided is whether Ken, in repairing the fan *in situ*, has done an unauthorised act or whether he was merely carrying out an authorised act in an unauthorised manner. The court would have to decide whether the authorised act was repairing the fan (that is, the wide approach to course of employment, as in *Rose v Plenty* and *Limpus*) or whether it was to repair the fan in the electrical workshop (that is, the narrow construction, as in *Conway v Wimpey* (1951)). It is submitted that a court would take the former approach and thus Lomad would be liable for Ken's negligence.

While liability seems likely, it should be noted that in recent years the courts have distinguished between careless and deliberate acts, and have taken a very narrow view of the course of employment where deliberate acts are concerned (see *Heasmans v Clarity Cleaning Ltd* (1987); *Irving v Post Office* (1987)). Perhaps the most dramatic example of this approach is to be found in *General Engineering Services v Kingston and St Andrews Corp* (1989). Here firemen who drove very slowly to the scene of a fire were held not to be within the course of their employment in so doing. They were employed to travel to the scene of the fire as quickly as reasonably possible and, in travelling as slowly as possible, they were not doing an authorised act in an unauthorised manner, rather they were doing an unauthorised act. However, one of the grounds for the decision in *General Engineering Services* included the finding that 'this decision [that is, the slow driving] was not in furtherance of their employer's business' (*per* Lord Ackner). In Lomad's case, we are told that the reason for Ken's action was to save time, that is, it was in furtherance of the employer's business. It is thus submitted that Ken's situation is legally distinguishable from that in *General Engineering Services*.

We should also consider the extent of the liability, as we are told that Martin is of a rather nervous disposition and that a normal person would not have suffered nearly as much harm. Fortunately for Martin, Ken and Lomad must take their victim as they find him.

The rule covering remoteness of damage for personal injury is that the defendant need only foresee some harm to the person (*The Wagon Mound (No 1)* (1961)). The extent of the injury is irrelevant, even if it was unforeseeable (*Dulieu v White* (1901); *Smith v Leech Brain* (1962)), that is, Ken and Lomad must take their victim as they find him and are liable for his injuries. See, for example, *Brice v Brown* (1984), where a plaintiff with a hysterical personality disorder recovered a substantial sum for the extremely bizarre behaviour she suffered following her witnessing an accident to her daughter.

Turning now to Ken's damage, Lomad, as his employer, is under a duty to provide Ken with safe equipment (*Smith v Baker* (1891)). It is no defence for Lomad to show that it purchased the screwdriver from a reputable supplier because, by s 1(1) of the Employers' Liability (Defective Equipment) Act 1969, where the defect is attributable to the fault of a third party, it is deemed to be attributable to negligence on the part of the employer. There is, of course, no problem in proving that a screwdriver is 'equipment' under the Employers' Liability (Defective Equipment) Act 1969, as the House of Lords has twice taken a wide approach to the meaning of this term (*Coltman v Bibby Tankers* (1988); *Knowles v Liverpool City Council* (1993)). Although this would appear to make life simple for Ken, it would have to be shown that the defect was attributable to the manufacturer. If the screwdriver was relatively new, and it can be shown that nothing has happened since it left the manufacturers to cause the defect, the defect can be attributed to the manufacturer (*Mason v Williams and Williams Ltd* (1955)). Thus, by s 1(1), it will be attributed to negligence on the part of the employer. However, if the screwdriver had been in use for some time, it may be difficult to show that the defect was due to fault on the part of the manufacturer (see *Evans v Triplex Glass* (1936)). As the duty to provide safe appliances is not an absolute one, but merely one to take reasonable care (see *Toronto Power v Paskwan* (1915)), the Employers' Liability (Defective Equipment) Act 1969 might not apply. If, for any reason, this is the case, then the situation is covered by *Davie v New Merton Board Mills* (1959) and Lomad will not be liable if it had not been negligent; for example, if it had purchased the screwdriver from a reputable supplier and the defect was not discoverable on reasonable examination, no liability will arise.

Notes

Question 17

Iambic plc owns some premises and decides to have the rather old fashioned central heating system replaced with a modern, efficient system. It engages Lead Ltd to carry out this work and Lead Ltd sends two plumbers to Iambic's premises. While the plumbers are working, one of them carelessly leaves a blowlamp running and the partition to an office catches fire. Jenny, who is working in the office, is burnt. Peter, who is an employee of Iambic plc, carelessly leaves a screwdriver on the floor of another office, and Katherine trips over it and twists her leg. In the ensuing commotion caused by these two accidents, an unknown thief enters the premises and steals a sheepskin coat belonging to Richard, another employee of Iambic plc. Richard kept his coat in a cupboard which was not provided with a lock.

Advise Jenny, Katherine and Richard.

Answer plan

This is a typical employers' liability question in the sense that while mostly involving employers' liability, it also requires a discussion of vicarious liability.

The following points need to be discussed:

- liability of Iambic plc for negligence of Lead Ltd;
- liability of Iambic plc for negligence of its employees;
- liability of Iambic plc for negligence of Peter; and
- employers' liability – limits on duty of care.

Answer

Considering Jenny first, we need to see against whom any cause of action might lie.

It is clear that the plumber himself owes Jenny a duty of care under normal *Donoghue v Stevenson* (1932) principles, in that he can reasonably foresee that any lack of care on his part may cause injury to Jenny. There is no need to apply the modern incremental formulation preferred by the House of Lords in *Caparo Industries plc v Dickman* (1990) and *Murphy v Brentwood District Council* (1990). Indeed, in *B v Islington Health Authority* (1991), at first instance, Potts J stated that in personal injury cases the duty of care remained as it was pre-*Caparo*, namely, the foresight of a reasonable person (as in *Donoghue*), a finding that does not appear to have been disturbed on appeal (1992). The plumber will be in breach of this duty if a reasonable plumber placed in his position would not have acted in this way (*Blyth v Birmingham Waterworks* (1856)), and it is submitted that a reasonable plumber would not have carelessly left a blowlamp running. This breach must have caused Jenny's injury, and the 'but for' test in *Cork v Kirby MacLean* (1952) proves the required causal connection. Additionally, the injury suffered by Jenny must not be too remote, in that it must be reasonably foreseeable (*The Wagon Mound (No 1)* (1961)). All that the plumber need foresee is some personal injury; he need foresee neither the extent (*Smith v Leech, Brain* (1962)) nor the exact manner in which the harm occurs (*Hughes v Lord Advocate* (1963)). All these criteria are satisfied, and so the plumber has been negligent as regards his conduct to Jenny.

As the plumber is an employee of Lead Ltd, Lead Ltd will be vicariously liable for any tort committed by the plumber in the course of his employment. As we are told that 'while the plumbers are working, one of them carelessly ...', it would seem that the plumber has been careless within the course of his employment. Also, the fact that the carelessness is gross and its consequences are obvious will not take the action outside the course of employment (*Century Insurance v Northern Ireland Road Transport Board* (1942)). Hence, Jenny could sue Lead Ltd in respect of her injury.

From the facts of the problem, there seems to be no reason for assuming that Lead Ltd is anything other than an independent contractor. The normal rule is that an employer is not liable for the torts committed by an independent contractor during the course of the contractor's duties (*Morgan v Girls Friendly Society* (1936); *D and F Estates v Church Commissioners* (1989)). There are some situations where liability will arise, namely, where the employer authorises the independent contractor to commit the tort (*Ellis v Sheffield Gas Consumers Co* (1853)); where he negligently chooses an incompetent contractor (*Pinn v Rew* (1961)); and where a non-delegable duty is imposed on him by common law (that is, a duty, the performance of which can be delegated but not the responsibility). The first two situations are not relevant here, but a non-delegable common law duty that may arise is that which exists where an independent contractor is employed to carry out work that is extra-hazardous (*Honeywill and Stein v Larkin Bros* (1934); *Alcock v Wraith* (1991)). In *Alcock*, the Court of Appeal held that a crucial question was: 'did the work involve some special risk, or was it from its very nature likely to cause damage?' It is suggested that plumbing does not satisfy these criteria; the use of a blowlamp may carry some special risk, but Iambic plc could claim that the plumber's negligence was merely collateral to the performance of his work and that, as an employer, it is not liable for this collateral negligence (*Padbury v Holliday and Greenwood* (1912)). If Jenny were to sue Iambic under the Occupiers' Liability Act 1957, she would be met with the defence in s 2(4)(b) that Iambic plc acted reasonably in entrusting the work to an independent contractor, took such steps as were reasonable to satisfy themselves that the contractor was competent and that the work had been properly done. As the work in question is technical, there would be no requirement for Iambic plc to check that it had been properly done (*Haseldine v Daw* (1941)).

If Jenny were to sue Iambic plc for breach of its common law duty as an employer to provide her with a safe place of work, she would be met with the defence that Iambic plc had taken reasonable steps to do so, as this duty is not absolute, but merely requires reasonable steps to be taken (*Latimer v AEC* (1953)). This situation should be distinguished from that in *McDermid v Nash Dredging and Reclamation* (1987), as in this case it was held to be no defence to breach of a non-delegable duty to show that the employer had delegated performance to a person, whether his employee or not, whom he reasonably believed to be competent to perform it (*per* Lord Brandon). In Iambic plc's case, it did not delegate the provision of a safe place of work to Lead Ltd or to the plumber.

Jenny is thus advised to sue Lead Ltd in respect of her injuries.

Turning now to Katherine, and following the analysis we used with the plumber, we can see that Peter has been negligent as regards his conduct to Katherine, as the necessary ingredients of duty, breach and damage are all present. We are told that Peter is an employee of Iambic plc and, assuming that when he left the screwdriver on the floor he was acting within the course of his employment, Iambic plc will be vicariously liable for his negligence. In addition to this secondary liability, Iambic plc, as Katherine's employer, has a primary duty to provide Katherine with a safe place of work. This is not an absolute duty, but merely requires Iambic plc to take reasonable steps to provide a safe place of work (*Latimer v AEC* (1953); see also *Gitsham v Pearce* (1991) for a more recent example). We need to decide therefore whether Iambic plc has taken such steps. Iambic plc has presumably instructed Peter not to leave any obstructions on the floor, but the problem for Iambic plc is that the duty to provide a safe place of work is non-delegable. In other words, an employer may entrust the performance of this duty to an employee, but he cannot thereby discharge his duty. In *McDermid v Nash Dredging and Reclamation*, Lord Brandon said: 'The essential characteristic of the (non-delegable) duty is that, if it is not performed, it is no defence for the employer to show that he delegated its performance to a person, whether his servant or not his servant, whom he reasonably believed to be competent to perform it. Despite such delegation, the employer is liable for the non-performance of the duty.' Thus, following *McDermid*, we can see that Iambic plc is in breach of its duty to provide a safe place of work.

Thus, Katherine is advised to sue Iambic plc for breach of its primary duty to provide a safe place of work, and as being vicariously liable for Peter's negligence.

Finally, we must consider Richard's situation. The courts have held consistently that the duty which an employer owes is a duty to safeguard the employee's physical safety (and this includes his mental state: *Walker v Northumberland County Council* (1994); *Hatton v Sutherland* (2002); *Barber v Somerset County Council* (2004)), but does not extend to protecting the economic welfare of the employee. This whole area was considered extensively in *Reid v Rush and Tomkins* (1990), where this distinction was upheld. In *Deyoung v Stenburn* (1946), in a similar fact situation, it was held by the Court of Appeal that no duty arose to protect the employee's clothing from theft (see also *Edwards v West Hertfordshire General Hospital Management Committee* (1957); *Crossley v Faithful & Gould Holdings Ltd* (2004)), hence Richard cannot sue Iambic plc for the loss of his coat. On the facts given, it seems most unlikely that he could sue either the plumber of Lead Ltd (as being vicariously liable) or Peter or Iambic plc (as being vicariously liable) for the loss of his coat, as such loss is not reasonably foreseeable and no duty of care would arise in respect of it.

This situation differs from that in *Stansbie v Troman* (1948), where a contractor left a house empty and the front door unlocked. The contractor was found liable for the subsequent theft of some property from the house, because in that situation it could be foreseen that a thief might enter and steal property from the house.

Notes

Question 18

To what extent, if any, does an employer's vicarious liability for the torts of his employees and his liability for breach of statutory duty add anything at all to the liability resulting from the employer's personal duty of care?

Answer plan

This is a question that requires careful thought. It would not be enough to merely list the main ingredients of vicarious liability, liability for breach of statutory duty and duty of care. What is required is a clear discussion of the limits of each of these doctrines, and the extent to which any remedies which are not available through the duty of care route can be supplemented by the other two routes and vice versa.

The following points need to be discussed:

- elements of the employer's personal duty of care, including an employee's physical and mental condition;
- vicarious liability of the employer, including limitations on vicarious liability; and
- statutory duties of the employer and rights of action in tort for breach.

Answer

The personal duty of care which an employer owes to his employee is to take reasonable care in all the circumstances for the employee's safety. Traditionally, this is formulated as the duty to provide competent fellow employees, properly maintained plant and equipment, and to provide a safe place and system of work (*Wilsons and Clyde Coal Co v English* (1938)). Two points should be noted immediately before these duties are considered in detail. First, these duties are non-delegable, by which we mean that an employer can delegate performance of these duties but by doing so he cannot thereby discharge those duties. In *McDermid v Nash Dredging and Reclamation* (1987), Lord Brandon stated that the essential duty of a non-delegable duty is that 'if it is not performed, it is no defence for the employer to show that he delegated performance to a person, whether his servant or not his servant, whom he reasonably believed to be competent to perform it. Despite such delegation, the employer is liable for the non-performance of the duty'.

Secondly, the duty that an employer owes to an employee is owed to that employee personally, with all his faults and idiosyncrasies, and is not a duty owed to his employees as an amorphous body (*Paris v Stepney Borough Council* (1951)).

Thus, in *Paris*, the employers were held to be in breach of their duty when they failed to provide an employee, who had sight in only one eye, with safety goggles. Although it was not usual practice to provide goggles for the work that the plaintiff carried out, the employers should have realised that in his particular case the consequences of an accident to his good eye would have been particularly disastrous. The criteria for breach were stated in some detail in *Stokes v Guest, Keen & Nettlefold* (1968) and these criteria were approved by the House of Lords in *Barber v Somerset County Council* (2004).

To consider these personal duties in turn, the first duty is to provide competent fellow employees. An example of this can be seen in *Hudson v Ridge Manufacturing* (1957), where it was held that an employer was liable for the consequences of a practical joke played on one employee by a fellow employee who was known to perpetrate such jokes over a considerable period of time. In such a situation, the employer might not be vicariously liable for the actions of the practical joker, as they might well not lie within the course of his employment. Consequently, the personal duty of care of the employer goes further than his vicarious liability. This restriction on the employer's vicarious liability has become particularly important in recent years, as the courts tend to take a much more restrictive approach to what constitutes the course of employment where the employee commits a deliberately wrongful act, as can be seen by the Court of Appeal decisions in *Heasmans v Clarity Cleaning* (1987) and *Irving v Post Office* (1987). In *Heasmans*, an employer was held not to be vicariously liable for the actions of an employee who was employed to clean telephones, but who made unauthorised telephone calls costing some £1,400. The court noted that the employee was employed to clean the telephones and that in using them he had done an unauthorised act which had taken him outside the course of his employment. In *Irving*, the employee, who worked for the Post Office and was employed to sort mail, wrote some racial abuse concerning the plaintiff upon a letter addressed to the plaintiff. The employee was authorised to write upon letters, but only for the purposes of ensuring that the mail was properly dealt with. It was held that the employers were not vicariously liable for the actions of the employee, as in writing racial abuse he was doing an unauthorised act and not an authorised act in an unauthorised manner. The court stated, *per* Fox LJ, that limits had to be set to the doctrine of vicarious liability, particularly where it was sought to make employers liable for the 'wilful wrongdoing' of an employee. Thus, in *General Engineering Services v Kingston and St Andrews Corp* (1989), the firemen who drove very slowly to a fire were held not to be in the course of their employment in so doing. They were employed to travel to the scene of fire as quickly as reasonably possible and, in travelling slowly, they were not doing an authorised act in an unauthorised manner, but were doing an unauthorised act.

In such situations, as the vicarious liability of the employer may be limited, the primary duty to take reasonable care for the employee's safety may provide the employee with a remedy.

The next personal duty of the employer is to provide properly maintained plant and equipment (*Smith v Baker* (1891)). This is a duty to do what is reasonably practicable (*Toronto Power Co v Paskwan* (1915)). However, by s 1(1) of the Employers' Liability (Defective Equipment) Act 1969, if an employee suffers personal injury in the course of his employment due to a defect in equipment provided by his employer, and the defect is due wholly or partly to the fault of a third party (whether identified or not), the injury is deemed to be also due to negligence on the part of the employer. Hence, an employer cannot discharge his duty to provide safe equipment by simply purchasing equipment from a reputable supplier, and his common law duty has been extended by statute from one which requires the exercise of reasonable care to what is, in effect, an absolute duty.

The employer is also under a duty to provide a safe place of work, again subject to taking steps that are reasonably practicable (*Latimer v AEC* (1953)). Similarly, the employer must not only provide such a system, but also ensure that it is operated safely (*McDermid v Nash Dredging and Reclamation* (1987)).

The above duties extend not only to safeguarding the employee's physical condition, but also to safeguarding the employee's mental state (*Walker v Northumberland County Council* (1994); *Hatton v Sutherland* (2002); *Barber*). However, in *Reid v Rush and Tomkins Group* (1990), it was held by the Court of Appeal that an employer had no duty to protect an employee's economic well being, although there was a duty to warn prospective employees of any physical risks inherent in a job (*White v Holbrook Precision Castings* (1985)).

An employer is also vicariously liable for the torts committed by an employee in the course of his employment. Thus, where by his negligence an employee injures a fellow employee, the employer will also be liable, providing that the act was in the course of the employment. There will often be an overlap between this ground of liability and that which arises from the employer's duty to provide competent fellow employees but, where a known practical joker causes damage to a fellow employee, as in *Hudson*, the joker may well be acting outside the course of his employment.

Finally, an employer may be liable for breach of statutory duty. As with the employers' personal duty of care, this is non-delegable, and an employee who relies on this cause of action will have to show that the breach conferred a right of action in tort. This usually gives rise to few problems as the statute will have been enacted for the benefit of a particular class of person, namely, employees (*Groves v Lord Wimborne* (1898)). However, a problem may arise in that the harm suffered may not be of the type that the statute was intended to prevent, that is, the *Gorris v Scott* (1874) situation. Thus, when considering a possible breach of s 14 of the Factories Act 1961, it has been held that the duty to fence dangerous machinery exists to protect the worker and to prevent him being injured through having his clothing caught in the machinery, but that is not relevant where a tool that a worker is using is caught in the machinery and the worker is thereby injured (*Sparrow v Fairey Aviation* (1964)). In such a case, however, there would be a clear breach of the employers' personal duty of care and indeed in *Fairey*, the breach of this duty was admitted, and the case was only brought to the House of Lords to determine whether or not there had been a breach of s 14.

Statutory duties may be subject to the 'reasonably practicable' requirement when they may add little to the common law duty of care or they may be absolute requirements. In the latter case, a breach is constituted when those requirements are not met, and the presence or absence of negligence is irrelevant.

Thus, to return to the statement under discussion, it can be seen that liability arising out of the employers' personal duty of care is often wider than vicarious liability or liability for breach of statutory duty. However, the three duties are separated and should not be confused. Obviously, where a statutory obligation is absolute, it will add to the employers' personal duty of care.

Notes

6 Product Liability

Introduction

The passing of the Consumer Protection Act 1987 and some recent important cases seemed to have jolted examiners' minds on this topic, and it is now commonly tested in examinations. Essay questions on the effect of the Consumer Protection Act 1987 and the differences that it has made on product liability are a favoured mode of testing this area but, where problem questions are set, the student must take care to consider the common law which has not been affected by the 1987 Act.

Checklist

Students must be familiar with the following areas:

(a) common law position:

- *dictum* in *Donoghue v Stevenson* (1932);
- intermediate examination;
- problems regarding defective product economic loss; and

(b) position under the Consumer Protection Act 1987:

- defects;
- persons liable;
- defences;
- loss caused;
- invalidity of exclusion clauses.

Question 19

'Although many manufacturers feared the introduction of the Consumer Protection Act 1987, they are in no a worse position since the Act than before.'

Critically discuss the above statement, paying particular attention to recent case law.

Answer plan

This question is typical of the essays which examiners are currently setting. It requires a discussion of the main elements of Pt 1 of the Act and a comparison of the statutory and common law regimes regarding product liability. The effect of the Act on manufacturers must be analysed, with particular reference to recent cases. Note, however, that the question requires candidates to discuss critically; it will not be sufficient merely to list the statutory requirements. These requirements must be compared with the still valid common law rules, and the effect of these changes from the manufacturers' point of view must be analysed.

In particular, the following points must be discussed:

- position under the Consumer Protection Act 1987;
- persons liable – s 2;
- definition of defect and guidelines for assessing safety – s 3;
- defences, especially the state of the art defence – s 4;
- limitations on property damage;
- position at common law;
- burden of proof; and
- requirement of causation and foreseeability.

Answer

Part 1 of the Consumer Protection Act 1987 was introduced into English law to implement the EC Directive 85/734/EEC relating to product liability. The main provision of the Act is to be found in s 2(1), which states that where any damage is caused wholly or partly by a defect in a product, the persons detailed in s 2(2) shall be liable for the damage. Section 2(2) lists the producer of the product, any person who holds himself out as the producer of the product, the importer of the product into the EC and (by s 2(3) in certain circumstances) the supplier of the product. A product is defined by s 1 as any goods or electricity, and includes a product which is comprised in another product.

The producer of the product is the manufacturer of the product, and a person holds himself out as being the producer if he puts his name or trade mark on the product or uses some other distinguishing mark, for example, a supermarket chain which sells its own brand products. The supplier of the product will be liable only where he is asked to identify the producer of the product and fails to do so.

By s 3, a product contains a defect where 'the safety of the product is not such as persons generally are entitled to expect'. Thus, the 1987 Act only requires the product to be reasonably safe; it does not impose a requirement of absolute safety. As almost any product is capable of being unsafe if misused (for example, a kitchen knife or an electric fire), the Act does not attempt to define safety, but instead provides a list of guidelines to be taken into account when considering what is meant by the term. So, by s 3(2), all the circumstances shall be taken into account, including: (a) the way and purposes for which the product has been marketed and any instructions and warnings provided; and (b) what might reasonably be expected to be done with the product. The Act also provides for certain defences, including the fact that the defect did not exist in the product at the relevant time, and the state of the art defence, namely, 'that the state of scientific and technical knowledge at the relevant time was not such that a producer of products of the same description as the product in question might be expected to have discovered the defect if it had existed in his products while they were under his control'.

This state of the art defence permitted by s 4(1)(e) is wider than that allowed in Art 7(e) of the original EC Directive, giving rise to the possibility that in appropriate cases a claimant could claim that a defendant could only rely on Art 7(e), and not on s 4(1)(e), as the true state of the art defence. However, in *EC Commission v UK* (1997), the European Court had to decide this very point, and held that in interpreting s 4(1)(e), there was nothing to suggest that the English courts would not arrive at the result that the Directive required. Indeed, *A v National Blood Authority* (2001) followed this case and adopted a purposive approach based on the Directive. In *A*, the court held that liability under the Directive is defect-based and that any question of fault by the manufacturer is irrelevant. Thus, a product is defective if it does not provide the level of safety a person is entitled to expect, whether or not that level of safety could have been achieved by the manufacturer. The Art 7(e) defence will not be relevant where there are known risks or risks which can reasonably be ascertained. Thus, in *Abouzaid v Mothercare (UK) Ltd* (2001), a manufacturer was held liable where a defect could easily have been discovered. However, in *Abouzaid*, although the manufacturer was found liable under the Consumer Protection Act 1987, he was not liable in common law negligence as he had acted as a reasonably prudent manufacturer when the product was made. These cases make it clear that a manufacturer is in a worse position under the Act than at common law. One small relief for manufacturers is that in *Richardson v LRC Products Ltd* (2001), which concerned an allegedly defective condom, it was stated that in determining what persons were entitled to expect under s 3, all the circumstances had to be taken into account, including any instructions or warnings, and the absence of any claim that the product was 100% effective. Thus, if a manufacturer makes no claim that his product is 100% effective, a failure does not necessarily prove the existence of a defect. However, in *Richardson*, it was reiterated that s 4 only affords a defence where the defect is one of which up to date scientific knowledge is ignorant.

One very important point which should be noted is that for all the defences contained within the Act, the burden lies on the defendant to prove the defence.

To establish liability under the Act, it is necessary to show that the defect caused the damage, either wholly or in part. There is no requirement of foreseeability; only causation need be shown.

Finally, it should be noted that the Act covers death or personal injury and damage to property. The property in question must be of the type which is normally intended for private use and which was intended for private use by the claimant. Consequently, damage to business property lies outside the scope of the Act. Damage to the product itself is excluded and there is a minimum value of £275 for property damage, below which damages cannot be awarded. By s 7, liability under the Act cannot be restricted or excluded.

To see to what extent this has changed the law, it is necessary to study the (still existing) common law. In *Donoghue v Stevenson* (1932), Lord Atkin stated: 'A manufacturer of products, which he sells in such a form as to show that he intends them to reach the ultimate consumer in the form in which they left him with no reasonable possibility of intermediate examination, and with the knowledge that the absence of reasonable care in the preparation or putting up of the products will result in an injury to the consumer's life or property, owes a duty to the consumer to take that reasonable care.'

The first problem that a consumer had was to identify the manufacturer. This may have been impossible, so that a donee of goods as opposed to a purchaser could be without a remedy.

Additionally, the remedy may have been, in practice, worthless where the manufacturer was based entirely outside the jurisdiction. In such a case, under the 1987 Act, the consumer could proceed against the importer of the goods into the EC.

The next hurdle that a consumer had to overcome was to show the absence of reasonable care on the part of the manufacturer. It could be difficult to show that the defect arose in manufacture, especially where the product had left the manufacturer's control some time previously. Thus, in *Evans v Triplex Safety Glass* (1936), where the owner of a car claimed that the windscreen was defective, he failed in his claim. The windscreen had been in use in the car for about one year and the plaintiff could not show that the defect in the glass was due to negligence on the part of the manufacturer. On the other hand, in *Mason v Williams and Williams* (1955), the plaintiff succeeded in proving that the manufacturers were negligent by showing that nothing had happened to the product after it left the manufacturers' possession that could have caused the defect.

This problem for the consumer remains under the Act, as s 4(1)(d) provides that it is a defence for the manufacturer to show that the defect did not exist in the product at the relevant time. Thus, causation remains a problem for the consumer, both at common law and under the 1987 Act, but unlike the common law there is no requirement of foreseeability of damage under the Act. Liability under the Act does not extend to damage caused to the product itself, whereas at common law recovery for defective product economic loss was allowed by the House of Lords in *Junior Books v Veitchi* (1983). However, *Junior Books* has been subject to intense judicial criticism, and later cases have tended to confine it within its specific facts. Thus, it was not followed in *Aswan Engineering Establishment v Lupdine* (1987); *Simaan General Contracting v Pilkington Glass* (1988); *Greater Nottingham Co-op v Cementation Piling and Foundations* (1989); or *D and F Estates v Church Commissioners* (1989).

Perhaps the view of the courts of *Junior Books* can best be summed up by some judicial statements of high authority. In *D and F Estates*, Lord Bridge stated that 'the consensus of judicial opinion seems to be ... that the decision cannot be regarded as laying down any principle of general application in the law of tort'. Likewise, Lord Oliver stated that it was 'really of no use as an authority on the general duty of care'. In *Simaan General Contracting*, Dillon LJ stated that *Junior Books* had been 'the subject of so much analysis and discussion with differing explanations of the basis of the case that the case cannot now be regarded as a useful pointer to any development of law ... indeed I find it difficult to see that future citation from *Junior Books* can ever serve any useful purpose'. In the light of these *dicta*, it came as no surprise when the High Court refused to follow *Junior Books* in *Nitrigin Eireann Teoranta v Inco Alloys* (1992), holding that it was 'unique' and hence no claimant could nowadays be advised to rely on *Junior Books*. Practically, therefore, in this respect the common law and the 1987 Act are identical.

Thus, it can be seen that it is incorrect to state that manufacturers are in no worse a position since the Act than before. Apart from the lower limit of £275 for property damage, below which an action may not be brought, the manufacturer is no better off than before as regards causation or defective product economic loss. Manufacturers are worse off in that a consumer has no requirements relating to foreseeability of damage, exclusion clauses are invalid and the defendant, that is, the manufacturer, will suffer the burden of any defence. In particular, it seems clear that recent cases have curtailed the extent of the state of the art defence in s 4(1)(e), so that manufacturers who would escape liability under common law negligence are now liable under the 1987 Act.

Notes

Question 20

Alice buys a toaster to give to her son, Bernard, who has just moved into a new flat. Because Bernard is having the flat decorated, he stores the toaster in a drawer in the kitchen and does not use it until the decorating is finished, some four weeks later. Due to the fact that it has been carelessly wired during manufacture, the toaster overheats and catches fire. Bernard suffers an electric shock when he attempts to put out the flames. His newly decorated kitchen is partially ruined and has to be re-papered. Also, the toaster is destroyed, together with a pocket dictaphone that Bernard uses in his job as a self-employed computer consultant.

Advise Bernard.

Answer plan

This problem calls for a discussion of Bernard's rights under the Consumer Protection Act 1987 and at common law.

The following aspects should be considered:

- persons liable under s 2(2) of the Consumer Protection Act 1987;
- criteria for existence of a defect – s 3(1) and (2);
- defences available under s 4 and burden of proof;
- any restrictions on property damage set by s 5 – minimum value, business property and defective product economic loss;
- common law action under *Donoghue v Stevenson* (1932) and differences between this action and the statutory remedy;
- need to show foreseeability of damage; and
- restrictions on type of damage recoverable.

Answer

Bernard should be advised of his rights under the Consumer Protection Act 1987 and at common law.

Under s 2(1) of the 1987 Act, the producer of the toaster will be liable for any damage caused by a defect in the toaster. Thus, the manufacturer of the toaster is liable, as is any person who holds himself out as the producer; for example, a shop who sold the toaster to Alice using its own brand name or trade mark or, if the product has been manufactured outside the EC, the importer into the EC of the product. If Bernard cannot identify any of these persons, he can ask the supplier of the toaster to identify such persons and, if the supplier fails to do so, the supplier will incur liability. Thus, Bernard should have no difficulty in identifying a potential defendant.

Next, Bernard must show that there was a defect in the product and that this defect caused the damage. A defect is defined by s 3(1) as existing if the safety of the product is not such as persons generally are entitled to expect. The 1987 Act does not require the product to be absolutely safe; it is enough that it is reasonably safe. The Act does not define safety as such, but instead gives a number of guidelines which are to be taken into account in determining whether or not the product is safe. By s 3(2), all the circumstances are to be taken into account, including the manner and purpose for which the product has been marketed, any instructions or warnings and what might reasonably be expected to be done with the product. As the product in question is a toaster and Bernard has used it for this purpose, and the toaster has overheated and caught fire because it has been carelessly wired during manufacture, it seems clear that it is unsafe and thus contains a defect. It seems clear also that this defect caused the damage which Bernard has suffered. Under the 1987 Act, it is sufficient for Bernard to prove causation and there is no requirement that the damage be reasonably foreseeable, so *prima facie* all the damage suffered is recoverable.

A possible defence for the producer is contained in s 4(1)(d), in that the defect did not exist in the product at the relevant time and, as Bernard kept the toaster for four weeks prior to using it, this defence must be considered. At common law, this is usually proved by showing that nothing happened to the product after it left the defendant's possession that could have caused the defect. Thus, in *Evans v Triplex Safety Glass* (1936), where the owner of a car claimed that the windscreen was defective, he failed in his claim. The windscreen had been in use in the car for about one year and the plaintiff could not show that the defect in the glass was due to negligence on the part of the manufacturers. On the other hand, in *Mason v Williams and Williams* (1955), the plaintiff succeeded in proving that the manufacturers were negligent by showing that nothing had happened to the product after it left the manufacturers' possession that could have caused the defect. As the time gap for Bernard is only four weeks and during that time the toaster lay in a drawer, it should not be difficult for Bernard to demonstrate that the defect arose in the manufacture and, in any event, the burden will be on the manufacturer to prove this defence (s 4(1)).

Thus, under the Act, Bernard can recover for the damage to his kitchen and for the electric shock he suffered. By s 5(4), he can only recover in respect of property damage if the damage exceeds £275, but that seems likely on the facts that we are given. As regards the dictaphone, it should be remembered that the Act is designed to benefit consumers. By s 5(3), liability does not arise in respect of property which is not obviously intended for private use or consumption, and which is not intended to be used by the claimant mainly for his own private use or consumption. The dictaphone does not satisfy both of these requirements, and so damage in respect of it cannot be recovered under the Act. Turning to the toaster itself, by s 5(2), damage to the product itself is excluded, so Bernard cannot claim for the damage to the toaster.

At common law, Bernard must rely on the *dictum* of Lord Atkin in *Donoghue v Stevenson* (1932) that: 'A manufacturer of products, which he sells in such a form as to show that he intends them to reach the ultimate consumer in the form in which they left him with no reasonable possibility of intermediate examination, and with the knowledge that the absence of reasonable care in the preparation or putting up of the products will result in an injury to the consumer's life or property, owes a duty to the consumer to take that reasonable care.' Thus, Bernard can proceed against the manufacturer of the toaster. Claimants have sometimes been allowed to proceed against suppliers, but that has been in cases where the supplier is under a duty to inspect the goods and fails to discharge this duty, for example, *Haseldine v Daw* (1941). A supplier or retailer of electrical goods would not have such a duty imposed on him, so Bernard could sue the manufacturer only.

As against the manufacturer, Bernard would have, on the facts given, little difficulty in establishing a breach of duty due to the presence of the defect. A reasonable manufacturer of toasters would not allow such a product into general circulation with such a defect (*Blyth v Birmingham Waterworks* (1856)). If the manufacturer were to claim that the defect did not exist in the toaster when it left his possession, he would be met with arguments similar to those discussed under the 1987 Act, although at common law the burden would lie on Bernard to prove the breach. Bernard must prove that the breach or defect caused his damage, and the 'but for' test in *Cork v Kirby MacLean* (1952) shows the required causal connection. Finally, Bernard must demonstrate that the damage which flowed from the breach was not too remote, in that it was reasonably foreseeable (*The Wagon Mound (No 1)* (1961)). Given the careless wiring, the damage which occurred is reasonably foreseeable. Thus, Bernard can recover for the damage to his kitchen and for the electric shock, and no minimum value will apply to the common law action for property damage. As regards the damage to the toaster, this is defective product economic loss. At common law, recovery for defective product economic loss was allowed by the House of Lords in *Junior Books v Veitchi* (1983). However, *Junior Books* has been subject to intense judicial criticism, and later cases have tended to confine it within its specific facts. Thus, it was not followed in *Aswan Engineering Establishment v Lupdine* (1987); *Simaan General Contracting v Pilkington Glass* (1988); *Greater Nottingham Co-op v Cementation Piling and Foundations* (1989); or *D and F Estates v Church Commissioners* (1989).

Perhaps the view of the courts to *Junior Books* can best be summed up by some judicial statement of high authority. In *D and F Estates*, Lord Bridge stated that 'the consensus of judicial opinion seems to be ... that the decision cannot be regarded as laying down any principle of general application in the law of tort'. Likewise, Lord Oliver stated that it was 'really of no use as an authority on the general duty of care'. In *Simaan General Contracting*, Dillon LJ stated that *Junior Books* had been 'the subject of so much analysis and discussion with differing explanations of the basis of the case that the case cannot now be regarded as a useful pointer to any development of law ... indeed I find it difficult to see that future citation from *Junior Books* can ever serve any useful purpose'. In view of this discussion, it should come as no surprise to learn that the High Court refused to follow *Junior Books* in *Nitrigin Eireann Teoranta v Inco Alloys* (1992), holding that it was 'unique'.

Thus, Bernard should be advised that there is little hope of a future court following *Junior Books,* and that the damage to the toaster is irrecoverable at tort. The damage to the dictaphone, however, is recoverable, for there is no common law requirement that the claimant be a consumer rather than a commercial user. Note that the phrase 'consumer' in Lord Atkin's judgment now means 'user' (*Mason v Williams and Williams* (1955)).

Notes

Question 21

Hilary and Janet work together. Janet agreed to cut and dye Hilary's hair one evening. After cutting Hilary's hair, Janet applied a dye which she bought from Blondie plc, who also manufactures the dye. After a few minutes, Hilary suffered an extremely painful allergic reaction to the dye and Janet washed the dye out. Several hours later, large portions of Hilary's hair fell out and her scalp turned bright red. As a result, Hilary cancelled a holiday that she was planning to take in Nepal, which cost £2,000.

Advise Hilary as to her legal rights.

Answer plan

This is a question on product liability which requires an analysis of the position of Blondie plc at common law and under statute regarding product liability, of Janet in negligence and any possible liability for the loss of the holiday.

The following points need to be discussed:

- Janet's liability to Hilary in negligence;
- liability of Blondie plc to Hilary at common law and under the Consumer Protection Act 1987;
- advantages of proceeding under the 1987 Act; and
- liability of Janet and Blondie plc in respect of the holiday.

Answer

Let us first consider any liability that Janet might have incurred to Hilary. Janet will owe a duty of care to Hilary under normal *Donoghue v Stevenson* (1932) principles, in that she can reasonably foresee that any lack of care on her part may cause injury to Hilary. There is no need to apply the modern incremental formulation of the test for the existence of a duty of care preferred by the House of Lords in *Caparo Industries plc v Dickman* (1990) and *Murphy v Brentwood District Council* (1990). Indeed, in *B v Islington Health Authority* (1991), at first instance, Potts J stated that in personal injury cases, the duty of care remains as it was pre-*Caparo*, namely, the foresight of a reasonable person (as in *Donoghue*), a finding that does not appear to have been disturbed on appeal (1992). We must next decide whether Janet is in breach of this duty, that is, that a reasonable person, or rather a reasonable hairdresser in Janet's position, would not have acted in this way (*Blyth v Birmingham Waterworks* (1856); *Bolam v Friern Hospital Management Committee* (1957)).

It is true that Janet is not a professional hairdresser but, as she has professed to have the skill of a hairdresser, she will be judged by the standard of a competent hairdresser. However, in *Philips v Whiteley* (1938), it was held that a jeweller who pierced ears for earrings was only under a duty to take the precautions which might reasonably be expected of a jeweller, and not meet the standards of cleanliness which would be expected of a surgeon. Similarly, in *Shakoor v Situ* (2000), it was held that a practitioner of Chinese herbal medicine should be judged by the reasonable standards of a reasonably careful practitioner of that art, rather than the standard of an orthodox medical practitioner. However, in *Wells v Cooper* (1958), it was held that a householder who did some work around the house must meet the standard of a reasonably competent carpenter. It is thus suggested that the earlier conclusion regarding Janet's duty of care is correct; her case is closer to *Wells* than *Philips*, in that in *Philips* there were two possible standards to apply, which did not exist in *Wells* or in Janet's case. We are not told whether or not the dye carried a warning regarding its application, for example, that a small test should be made before general use, and if Janet disregarded any such warning she will be in breach of her duty. In any event, it is submitted that a reasonably competent hairdresser (the standard by which Janet must be judged) would be aware that some persons might be particularly sensitive to hair dyes and would carry out a preliminary test. As Janet has apparently not done this, she is in breach of her duty. Finally, it must be shown that this breach caused Hilary's injuries, and the 'but for' test in *Cork v Kirby MacLean* (1952) proves the required causal connection. In addition, the damage suffered by Hilary must not be too remote, that is, it must be reasonably foreseeable (*The Wagon Mound (No 1)* (1961)). Certainly, some allergic reaction is foreseeable if no pre-testing is carried out, and this will be sufficient to found liability for the painful reaction, the loss of hair and the discoloured scalp. There is no need for Janet to foresee the extent of the injuries suffered by Hilary, as the rule with personal injuries is that the defendant need only foresee the kind of injuries, not the extent (*Smith v Leech Brain* (1962)). Nor would it be any defence for Janet to show that Hilary had particularly sensitive skin, as a tortfeasor must take his victim as he finds him (the egg shell skull rule) (*Dulieu v White* (1901); *Smith v Leech Brain* (1962)). A problem may arise as regards Hilary's cancelled holiday and the loss that entails – *prima facie*, this is not a reasonably foreseeable consequence of Janet's negligence and would be recoverable only if it was so, for example, because Hilary had told Janet that she had arranged a holiday in the near future. In that case, the cancelled holiday would be reasonably foreseeable and hence recoverable.

There is, however, a particular problem to suing Janet, which is that she may not be able to satisfy judgment, so we need to consider whether Blondie plc is liable to Hilary. Considering first the common law situation, it was established in *Donoghue v Stevenson* (1932) that 'A manufacturer of products, which he sells in such a form as to show that he intends them to reach the ultimate consumer in the form in which they left him with no reasonable possibility of intermediate examination, and with the knowledge that the absence of reasonable care in the preparation or putting up of the products will result in an injury to the consumer's life or property, owes a duty to the consumer to take reasonable care' (*per* Lord Atkin). This liability has been extended to include suppliers as well as manufacturers, so it will apply to Blondie plc whether it manufactured the dye or merely supplied it to Janet. A hurdle which needs to be overcome in holding Blondie plc liable is that common law liability will only arise where there is 'no real possibility of intermediate examination'. In *Kubach v Hollands* (1937), it was held that the presence of an adequate warning was enough to discharge this duty, so if the dye bottle supplied by Blondie plc carried a suitable warning, this would exempt it from liability. If a suitable warning was not provided, Blondie plc could argue that the true cause of Hilary's injuries was not its breach, but rather the negligent act of Janet in not carrying out a pre-test as a reasonably competent hairdresser would have been expected to, that is, that Janet's negligent act is a *novus actus interveniens* which broke the chain of causation. Where it is alleged that the act of a third party, over whom the defendant has no control, has broken the chain of causation, it must be shown that the act was 'something unwarrantable, a new cause which disturbs the sequence of events, something which can be described as either unreasonable or extraneous or extrinsic' (*per* Lord Wright in *The Oropesa* (1943)). Therefore, the defendant will remain liable if the act of the third party is not truly independent of his negligence. In *Knightley v Johns* (1982), the Court of Appeal held that negligent conduct was more likely to break the chain of causation than non-negligent conduct. In *Knightley,* there were so many errors and departures from common sense procedures that the chain of causation had been broken. We have already decided, in considering Janet's possible liability, that her actions in not carrying out a pre-test were negligent, and it is submitted that this negligent act was such a departure from common sense procedures as to break the chain of causation, and relieve Blondie plc of liability. The extent of Blondie plc's liability, should it exist at common law, will be governed by the reasonably foreseeable criterion, so it would not be liable for any loss as regards Hilary's aborted holiday.

We next need to consider if any liability arises under the Consumer Protection Act 1987. By s 2(1) of this Act, where any damage is caused wholly or partly by a defect in a product, certain persons are liable for the damage. Those persons are the producer of the product, who we are told is Blondie plc. Next, Hilary will have to show that the product contained a defect, in that its safety was not such as persons generally are entitled to expect (s 3(1)). The Act does not attempt to define safety, but requires all the circumstances to be taken into account, including any instructions or warnings provided (s 3(2)). Thus, similar considerations will apply as in the earlier discussion regarding common law liability, and a similar defence of *novus actus interveniens* will be available to Blondie plc, as s 2(1) expressly requires causation to be proved. One advantage that accrues to Hilary in proceeding under the Act rather than at common law is that there is no requirement of foreseeability under the Act, and if Blondie plc were to be found liable, all the harm suffered by Hilary is recoverable. If Blondie plc wishes to raise any of the defences open to it under s 4, it will have to prove these defences. The only relevant defence appears to be s 4(1)(e), that the state of scientific and technical knowledge at the relevant time was not such that a producer of products of the same description as the product in question might be expected to have discovered the defect if it had existed in his products while they were under his control. However, given the universal and sophisticated testing of hair products for allergic responses, this defence seems unlikely to succeed. Furthermore, in *A v National Blood Authority* (2001), the High Court held that liability under the 1987 Act is defect-based, and it is not necessary to prove a fault on the part of the manufacturer. Thus, a product is defective if it does not provide the expected level of safety, whether or not the manufacturer could have avoided that lack of safety. A similar conclusion was reached in *Abouzaid v Mothercare (UK) Ltd* (2001). Also, in *Richardson v LRC Products Ltd* (2000), it was stated that s 4 will only provide a defence where the defect is one of which up to date scientific knowledge is unaware.

There is also a minimum value to actions under the Act, but this only applies to property damage and Hilary's action is a personal injuries one.

Overall, therefore, it seems that Hilary has a good case against Janet, but that the chances of success against Blondie plc are more problematic. Subject to the *novus actus* point raised above, Blondie plc may also incur liability under the General Product Safety

Regulations 1994 (SI 1994/2328). However, these Regulations, which are similar to the 1987 Act, provide only for criminal penalties on breach and so are only of limited use to Hilary.

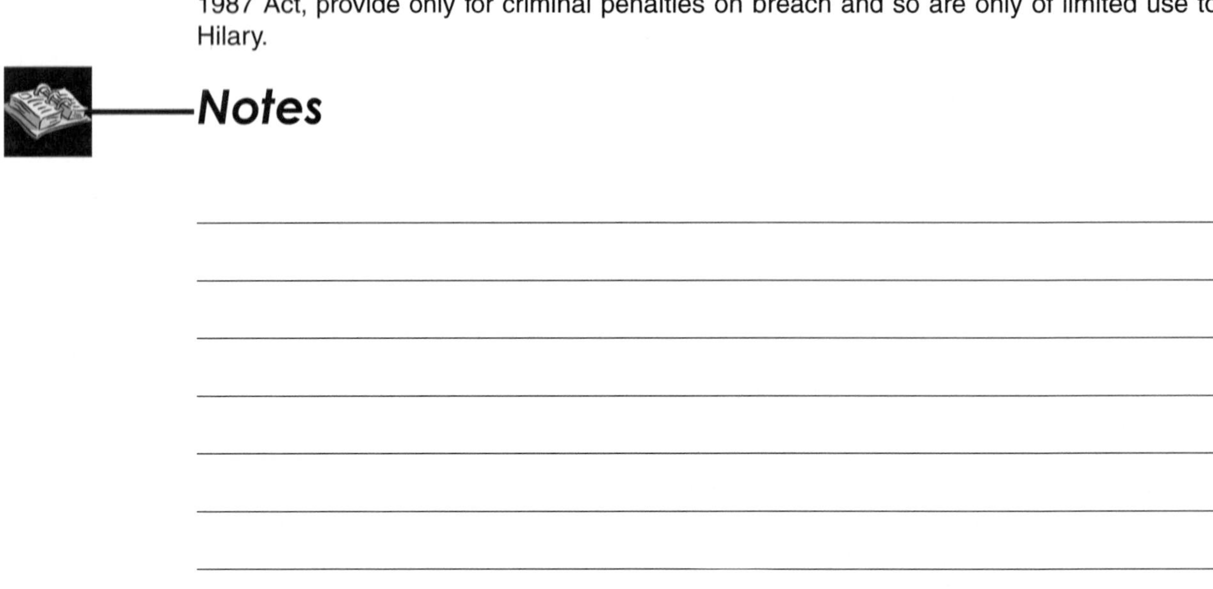

7 Occupiers' Liability

Introduction

Occupiers' liability is a specialised branch of the tort of negligence and is tested in most examinations year after year. The area is governed by statute, namely the Occupiers' Liability Act 1957 and the Occupiers' Liability Act 1984. Thus, in addition to the common law concepts of duty, breach, causation and remoteness, attention must be paid to the statutes and the exact words used therein.

The 1984 Act is also frequently tested. Be aware that in moving around premises, a person may well change in status from a visitor to a non-visitor, that is, from being subject to the 1957 Act to being subject to the 1984 Act.

Checklist

Students must be familiar with the following areas:

(a) who are occupiers, visitors and non-visitors;

(b) duty regarding children, warnings and independent contractors;

(c) exclusion of duty; and

(d) circumstances under which a duty to a non-visitor arises and nature of this duty.

Question 22

Arthur inherits a large and dilapidated house from his mother. He moves in and decides to have substantial renovations carried out by Askew Alterations Ltd, a local company which specialises in renovating old property. Whilst these alterations are in progress, Arthur decides to hold a party to welcome his new neighbours.

Basil and his five year old daughter Clara attend. Clara becomes bored and wanders into a room marked 'Danger – do not enter' and is injured. Basil, while looking for Clara in that room, turns on a light switch that has not been completely finished and suffers an electrical shock.

Cedric, who is aware that Arthur's mother kept a good wine cellar, goes down to the cellar intending to help himself to some wine, but slips on a cork on the steps and breaks both his legs.

Advise Basil, Clara and Cedric.

Answer plan

This is a standard occupiers' liability question, in that it involves the areas of independent contractors, children and visitors becoming non-visitors.

The following points need to be discussed:

- occupiers and visitors;
- Arthur's duty to visitors generally;
- Arthur's duty to Clara;
- effect of warning notice; and
- Arthur's duty to Cedric.

Answer

Arthur is the occupier of his house, as he has 'sufficient control over the premises that he ought to realise that any failure on his part to use care may result in injury to a person coming lawfully there' (*Wheat v Lacon* (1966), *per* Lord Denning). Basil, Clara and Cedric are Arthur's visitors (*Wheat*), and we must ascertain the nature of the duty Arthur owes to each of his visitors and decide whether he is in breach of that duty.

Arthur owes each of his visitors the common duty of care (s 2(1) of the Occupiers' Liability Act 1957), and this duty is to take such care as in all the circumstances is reasonable to see that the visitor will be reasonably safe in using the premises for the purposes for which he is invited or permitted by the occupier to be there. It should be noted that it is the visitor who must be reasonably safe and not the premises (see, for example, *Ferguson v Welsh* (1987)). Thus, the fact that repairs are being carried out to Arthur's house, which is in a dilapidated condition, does not (without more) constitute a breach of duty.

Turning now to Clara, by s 2(3)(a) of the 1957 Act, Arthur must be prepared for children to be less careful than adults. In *Latham v Johnson and Nephew Ltd* (1913), Lord Hamilton stated that there may be a duty not to lead children into temptation. Having said this, if the danger is obvious, even to a child, the occupier will not be liable (*Liddle v Yorkshire (North Riding) County Council* (1934)). With very young children of course almost anything can be a danger, but here an occupier will be able to rely on the decision of Devlin J in *Phipps v Rochester Corp* (1955). It was held that reasonable parents would not allow small children to go unaccompanied to places which may be unsafe for them, that both parents and occupiers must act reasonably and each is entitled to assume that the other has so acted.

Considering all the circumstances of the case, it should have been clear to Basil that Arthur's house was in the process of redecoration and would hence contain danger that might not be obvious to a small child. Following *Phipps*, it would seem that Basil has not acted reasonably and that Arthur was justified in relying on Basil to look after Clara. Arthur could also argue that the sign 'Danger – do not enter' is a warning which discharges his duty under s 2(4)(a), but to achieve this the warning must in all the circumstances be enough to enable the visitor to be reasonably safe. The sign does not seem to be a warning at all, in that it makes no attempt to describe the danger. It is rather a prohibition on the spatial extent to which the visitor is entitled to be on the premises, which the occupier is entitled to do (*The Calgarth* (1927)). In any event, it would not be enough in all the circumstances to discharge the duty owed to a small child, nor to turn the small child into a non-visitor.

In view of the decision of the House of Lords in *Edwards v Railway Executive* (1952), it is most unlikely that the court would imply a licence in favour of Clara, turning her into a visitor – see the judgment of Lord Goddard.

Hence, our advice to Clara is that she cannot sue Arthur, but that she could sue Basil in negligence. Basil owes Clara a duty of care under normal *Donoghue v Stevenson* (1932) principles. As a duty of care has previously been established, there is no need to proceed to the modern incremental formulation of a duty of care that was preferred by the House of Lords in *Caparo Industries plc v Dickman* (1990) and *Murphy v Brentwood District Council* (1990). Basil is in breach of his duty in allowing Clara to wander off in a house that was still being renovated, as this would not have been the action of a reasonable parent placed in Basil's position (*Blyth v Birmingham Waterworks* (1856)). Finally, the harm that Clara has suffered was caused by Basil's breach of duty, as shown by applying the 'but for' test of Lord Denning in *Cork v Kirby MacLean* (1952). This damage was also reasonably foreseeable, as required by *The Wagon Mound (No 1)* (1961).

We have discussed the duty Arthur owes to Basil and must now consider whether Arthur is in breach of this duty. *Prima facie*, it is a breach of duty to allow persons to come into contact with unsafe light switches, but Arthur has attempted to discharge his duty via the notice. The notice seems to be insufficient as a warning notice, as it does not describe the nature of the danger in any way, and so by s 2(4)(a) would not be enough to make the visitor reasonably safe. In *Rae v Mars UK* (1989), it was held that where an unusual danger exists, the visitor should not only be warned, but a barrier or additional notice should be placed to show the immediacy of the danger. Arthur has not complied with this condition and, therefore, it seems that the notice is insufficient as a warning notice. However, Arthur could rely on s 2(4)(b). This section states that where damage is caused to a visitor by a danger due to the faulty execution of any work of construction, maintenance or repair by an independent contractor employed by the occupier, the occupier is not to be treated, without more, as answerable for the danger. This applies if, in all the circumstances, the occupier had acted reasonably in entrusting the work to an independent contractor and had taken such steps (if any) as he reasonably ought in order to satisfy himself that the contractor was competent and that the work had been properly done.

The renovation is covered by s 2(4)(b), as it is reasonable to entrust it to independent contractors, and as we are told that Askew Alterations specialises in renovating old property, it would seem that it is competent. The question, therefore, is what (if any) steps Arthur ought reasonably to have taken to satisfy himself that the work had been properly done. Despite the words 'had been properly done', it was held by the House of Lords in *Ferguson v Welsh* (1987) that it could apply where the work was still being done and had not been completed. The rule is that the more technical the work, the less reasonable it is to require the occupier to check it (*Haseldine v Daw* (1941); *Woodward v Mayor of Hastings* (1945)). In Arthur's circumstances, it would seem that as the work is technical, there is no requirement to check that the work had been properly done, and Arthur has discharged his duty by employing reasonable independent contractors.

One could consider whether Askew Alterations is also an occupier. Although Arthur is an occupier, an independent contractor may also be an occupier, for control need not be exclusive (*Wheat v Lacon* (1966)). The question that has to be decided is whether the independent contractors have sufficient control as in *AMF International v Magnet Bowling* (1968). In the present case, although we are told that renovations are continuing, it seems unlikely that Arthur would hold a party while the contractors are physically present and working. Although physical possession is not a necessary ingredient of control (*Harris v Birkenhead Corp* (1976)), it would seem that at the relevant time Askew Alterations was not in sufficient control to make it an occupier.

Hence, our advice to Basil is that he cannot sue Arthur, but that he could sue Askew Alterations in negligence. Askew Alterations owes Basil a duty of care under normal *Donoghue* principles. Also, as a duty of care has previously been established, there is no need to proceed to the modern incremental formulation of a duty of care that was preferred by the House of Lords in *Caparo* and *Murphy*. Askew Alterations is in breach of its duty in leaving the switch in an unsafe condition, as this would not have been the action of a reasonable electrician placed in Askew Alterations' position (*Blyth v Birmingham Waterworks* (1856)). Finally, the damage that Basil has suffered was caused by Askew Alterations' breach of duty, as shown by applying the 'but for' test of Lord Denning in *Cork v Kirby MacLean* (1952), and this damage was reasonably foreseeable, as required by *The Wagon Mound (No 1)* (1961).

Finally, we must consider advising Cedric. Although Cedric was initially a visitor, on entering the cellar, he became a non-visitor. An occupier may place a spatial limitation on the visitor's permission to enter (*The Calgarth* (1927)), but in such a case the limitation must be brought to the visitor's attention (*Gould v McAuliffe* (1941)). By implication, Cedric must have known that he did not have Arthur's permission to enter his wine cellar. Consequently, Cedric became a trespasser when he entered that part of the premises. Even if Cedric had had permission to visit the wine cellar, as he went there to steal Arthur's wine, Cedric must have known that he was entering the cellar in excess of the permission given to him and consequently became a trespasser.

Any duty now owed to Cedric is governed by the Occupiers' Liability Act 1984. By s 1(2) of the 1984 Act, an occupier will only owe a duty to a non-visitor if the occupier:

(a) is aware of the danger or has reasonable grounds to believe it exists;

(b) knows or has reasonable grounds to believe that the non-visitor is in the vicinity of the danger or may come into the vicinity; and

(c) the risk is one against which, in all the circumstances, he may reasonably be expected to offer the non-visitor some protection.

It seems unlikely that requirement (a) is satisfied as actual knowledge of the danger is required: *Ratcliff v McConnell* (1999); *Donoghue v Folkestone Properties Ltd* (2003). Requirement (b) is not satisfied, as Arthur had no reason to suspect that his guests would steal his property. Hence, no duty arises in respect of Cedric's accident in the cellar. See, for example, *Tomlinson v Congleton Borough Council* (2003); *Donoghue*. In addition, Arthur would have the defence of *ex turpi causa non oritur actio* (*National Coal Board v England* (1954)). Cedric's claim is based directly on the illegality, and is not merely incidental, to use the test preferred by the majority in the Court of Appeal in *Pitts v Hunt* (1990) and by the House of Lords in *Tinsley v Milligan* (1993). In *Revill v Newberry* (1996), a trespasser was allowed to recover, subject to a large deduction for contributory negligence, when he was injured during his criminal activity. The Court of Appeal stated that 'an occupier cannot treat a burglar as an outlaw'. As Arthur has not treated Cedric in this way, the *ex turpi* defence will be valid.

Notes

Question 23

Eric owns a waxwork museum in the seaside town of Westsea. He decides to have a new air conditioning system installed in the museum by Coolit plc, but Coolit can only carry out this work at the height of the tourist season. Rather than delay the job until winter or shut down while the work is being done and lose income, Eric decides to allow the public into the museum while the new air conditioning system is being installed. He places notices around the museum stating 'Danger – Work in Progress'. While the employees of Coolit are working in one part of the museum, some scaffolding which they have erected in another part of the museum collapses. The scaffolding injures Florence, who paid to enter the museum, and George, who entered without paying through the open back door that Coolit's employees were using to bring in equipment.

Advise Florence, who has suffered a fractured skull and had her spectacles broken, and George, who has suffered a broken shoulder and has had his new suit ruined.

Answer plan

This is a straightforward occupiers' liability question involving two occupiers, an independent contractor, a visitor and a non-visitor.

The following points need to be discussed:

- likelihood of Eric and Coolit both being occupiers;
- applicability of s 2(4)(b) of the Occupiers' Liability Act 1957 to Eric;
- liability of occupier to non-visitor; and
- damages recoverable.

Answer

Eric is the occupier of the museum as he has, *per* Lord Denning, sufficient control over the premises that he ought to realise that any failure on his part to use care may result in injury to a person coming lawfully there (*Wheat v Lacon* (1966)). In addition, Coolit may also be an occupier of the museum, for there is no need for control to be exclusive (*Wheat*). The question is whether Coolit, as an independent contractor, has sufficient control as in *AMF International v Magnet Bowling* (1968). This is of course a question of fact and, as presumably the installation of an air conditioning system in a waxworks museum would involve extensive work, Coolit may well be held to be an occupier and so also be liable together with Eric.

Florence is Eric's visitor (*Wheat*). Eric owes Florence the common duty of care by s 2(1) of the Occupiers' Liability Act 1957. By s 2(2), the duty is to take such care as in all the circumstances of the case is reasonable to see that the visitor will be reasonably safe in using the premises for the purposes for which he is invited or permitted to be there. It should be noted that it is the visitor who must be reasonably safe and not the premises, so the fact that renovations are taking place does not, by itself, constitute a breach of duty. The question as to whether the occupier is in breach of his duty is always a question of fact depending on the exact circumstances of the case. This can be seen by the differing decisions of the court in the factually similar cases of *Murphy v Bradford Metropolitan Council* (1992) and *Gitsham v Pearce* (1992).

Eric will, however, seek to rely on the defence contained in s 2(4)(b) of the 1957 Act, namely that where damage is caused to a visitor by a danger due to the faulty execution of any work of construction, maintenance or repair by an independent contractor employed by the occupier, the occupier is not to be treated without more as answerable for the danger. This applies if, in all the circumstances, the occupier acted reasonably in entrusting the work to an independent contractor and had taken such steps (if any) as he reasonably ought in order to satisfy himself that the contractor was competent and that the work had been properly done.

There is nothing in the facts of the problem to suggest that Coolit is anything but competent. The question, therefore, is what (if any) steps Eric ought reasonably to have taken to satisfy himself that the work had been properly done. Despite the words 'had been properly done' in the sub-section, it was held in *Ferguson v Welsh* (1987) that the obligation could arise where the work was still being done and had not been completed. The rule is that the more technical the work, the less reasonable it is to require the occupier to check the work (cf *Haseldine v Daw* (1941) and *Woodward v The Mayor of Hastings* (1945)).

The work of installing an air conditioning system is certainly technical, but we are told that the damage was caused by scaffolding collapsing. If this danger was obvious to a reasonable observer, Eric should have been aware of the danger and cannot bring himself within s 2(4)(b). However, if the careless work of Coolit was not apparent upon such examination, Eric will not be in breach.

Next, we need to consider whether the notices stating 'Danger – Work in Progress' discharge any duty owed by the occupier. By s 2(4)(a) of the 1957 Act, where damage is caused to a visitor by a danger of which he has been warned by the occupier, the warning is not to be treated without more as absolving the occupier from liability, unless in all the circumstances it was enough to enable the visitor to be reasonably safe. The notices do not indicate the nature of the danger, and it is a question of fact whether they were enough to allow a visitor to be reasonably safe. In *Rae v Mars UK* (1989), it was held that where an unusual danger exists, the visitor should not only be warned but a barrier or additional notice should be placed to show the immediacy of the danger. On the facts of the present case this had not been done. As the scaffolding fell on Florence, it seems that in addition to the notice, the area should have been roped off to keep visitors away from any possible danger.

Hence, our advice to Florence is that Eric may be able to avail himself of this statutory defence, so she would be better advised to sue Coolit under the Occupiers' Liability Act 1957 as occupiers and/or in negligence. Florence can recover for both the injury to her person and the damage to her property: s 1(3)(b) of the 1957 Act.

George is clearly not a lawful visitor of Eric or Coolit. George comes within the definition of a trespasser as laid down by Lord Dunedin in *Addie v Dumbreck* (1929) as a person who goes onto land without invitation and whose presence is either unknown to the proprietor or, if known, is objected to. The duty owed to George is covered by the Occupiers' Liability Act 1984. By s 1(3) of this Act, an occupier will only owe a duty to a non-visitor if:

(a) he is aware of the danger or has reasonable grounds to believe that it exists;
(b) he knows or has reasonable grounds to believe that the non-visitor is in the vicinity of the danger or may come into the vicinity; and
(c) the risk is one against which, in all the circumstances, he may reasonably be expected to offer the non-visitor some protection.

If the duty does arise then, by s 1(4), an occupier is to take such care as is reasonable in all the circumstances to see that the non-visitor does not suffer injury.

In George's case, requirement (a) is satisfied as Eric was aware of the danger (as can be shown by his placing of the warning notices: *Woollins v British Celanese* (1966)). Requirement (b) is not satisfied, as Eric has no reason to anticipate George's presence. In *White v St Albans City and District Council* (1990), it was argued that the very presence of a warning notice showed that the occupier had reason to suspect someone was likely to come into the vicinity of the danger, but this was rejected by the Court of Appeal. In both *Donoghue v Folkestone Properties Ltd* (2003; CA) and *Tomlinson v Congleton Borough Council* (2003; HL), the position of trespassers was considered and it was held in both cases that there was no duty owed in respect of obvious dangers. It was also held in both cases that no duty was owed where persons freely and voluntarily undertook an activity which involved some risk. In George's case, it could be argued that the danger from the activity involving scaffolding is obvious, and that using the back door to enter the premises involved an element of risk in that any warnings would not be apparent to George.

Even if the duty does arise, it can be discharged by a reasonable warning: s 1(5) (but see above discussion). By s 1(9), injury only includes personal injury and damage to property is expressly excluded (s 1(8)). Thus, in the unlikely event of George being able to establish liability under the 1984 Act, the damage to his suit would be irrecoverable.

The question that arises is if George could establish any liability, whether he would be met by the *ex turpi causa non oritur actio* defence as in *National Coal Board v England* (1954). The scope of this defence is difficult to ascertain from the decided cases. In *Euro-Diam v Bathurst* (1988), it was said that the defence rests on a public policy that the courts will not assist a claimant who has been guilty of illegal or immoral conduct of which the court should take notice, and the defence will apply if it would be an affront to the public conscience to grant the claimant relief. This test was used in *Thackwell v Barclays Bank* (1986); *Saunders v Edwards* (1987); and by Beldam LJ in *Pitts v Hunt* (1990). However, in *Pitts,* Dillon and Balcombe LJJ preferred to base their decisions on whether the plaintiff's claim was based directly on his illegal conduct or whether the illegal conduct was merely incidental. More recently, the House of Lords in *Tinsley v Milligan* (1993) rejected the affront to public conscience test. The test would now seem to be whether the claim is based directly on the illegal conduct (*Vellino v Chief Constable of Greater Manchester* (2002)). In George's case, it would seem that the *ex turpi* defence would succeed, so George should be advised that his chances of any recovery are extremely slim.

Notes

Question 24

'Any sensible occupier will exclude the onerous duty of care that he owes to both visitors and non-visitors as regards his occupation of premises.'

Discuss the above statement.

Answer plan

This question requires a discussion of the standard of care required of an occupier of premises and an assessment as to whether this duty is onerous, together with an assessment of the extent to which an occupier is free to exclude this duty. Particular care should be taken to discuss the recent cases involving non-visitors.

The following points need to be discussed:

- duty of care imposed on occupiers in respect of lawful visitors:
    - generally;
    - in specific circumstances;
    - extent to which this duty may be excluded; and
- duty of care imposed in respect of non-visitors, with especial reference to *Donoghue v Folkestone Properties* and *Tomlinson v Congleton Borough Council* and the extent to which it may be excluded.

Answer

We shall first consider the duty of care that an occupier of premises owes to his lawful visitors. This area is governed by the Occupiers' Liability Act 1957 which, by s 1(1), replaced the previous common law rules. By s 2(1), an occupier of premises owes the common duty of care to his visitors, except insofar as he is free to and does extend, restrict, modify or exclude his duty by agreement or otherwise. This common duty of care is defined in s 2(2) as the duty to take such care as in all the circumstances of the case is reasonable to see that the visitor is reasonably safe in using the premises for the purposes for which he is invited or permitted by the occupier to be there. It follows from s 2(2) that it is the visitor that must be reasonably safe and not the premises – an occupier may maintain his premises in an unsafe state, providing only that his visitors are safe. The duty contained within the Act is similar to the duty in a common law negligence action, as can be seen by a consideration of those cases which have decided whether or not there has been a breach of s 2(2) (see, for example, *Bell v Department of Health and Social Security* (1989); *Murphy v Bradford Metropolitan Council* (1992)). Whether or not this is an onerous duty is in many ways a subjective decision: doubtless, those persons who own a large number of properties might consider any duty owed in respect of those premises to be onerous, but it is submitted that a duty which goes no further than the standard *Donoghue v Stevenson* (1932) duty of care is not onerous.

One aspect in which the 1957 Act differs from the common law is that it makes specific provision for certain situations. Thus, s 2(3)(a) states that an occupier must be prepared for children to be less careful than adults. However, this would seem to add little to the *Donoghue* standard of care, for, although an occupier must not place temptation in children's way (*Latham v Johnson and Nephew Ltd* (1913); *Glasgow Corp v Taylor* (1922)), it has been held that an occupier will not be liable for dangers which are obvious even to children (*Liddle v Yorkshire (North Riding) County Council* (1934)). Thus, as regards children of 'tender years', the occupier is entitled to assume that reasonable parents will not allow such children to be in dangerous situations without protection and that parents will act reasonably. However, where an occupier does allow an allurement or trap to remain on his premises, the occupier will be liable for even small risks. Thus, in *Jolley v Sutton Borough Council* (2000), the House of Lords held that the ingenuity of children in finding ways of doing mischief to themselves or others should never be underestimated. The House overruled the finding of the Court of Appeal (1998) that the damage was not reasonably foreseeable, and it seems that a wide approach should be taken to reasonable foreseeability where children are concerned. As we have already suggested that a duty which goes no further than *Donoghue*, that is, which depends on the foresight of a reasonable person and requires that person to act reasonably, is not onerous, s 2(3)(a) would not represent an onerous extension of the occupiers' duty. Section 2(3)(b) provides that an occupier may expect that a person, in the exercise of his calling, will appreciate and guard against any special risks ordinarily incident to it. Hence, an occupier may employ a person to carry out a hazardous activity on his premises, and rely on this provision where that activity lies within that person's calling. However, this does not mean that merely because a visitor possesses a particular skill, that fact, in itself, is enough to discharge the duty of care owed to that visitor. Thus, in *Salmon v Seafarer Restaurants* (1983) and *Ogwo v Taylor* (1988), it was held that an occupier owes the same duty of care to a fireman as to any other visitor, and the question to be decided in all these cases was whether the injury to the visitor was reasonably foreseeable. An occupier is, of course, entitled to assume that the fireman will follow standard practice in fighting the fire. Again, this provision does not appear to impose an onerous duty on an occupier, as it too adds little to the *Donoghue* type of duty.

Another situation for which the 1957 Act makes specific provision is that it enables the occupier to discharge his duty by a warning. Section 2(4)(a) states that where a visitor has been warned of a danger, that warning will not of itself be enough to discharge the occupiers' duty of care unless in all the circumstances it was enough to allow the visitor to be reasonably safe. Thus, an occupier may discharge his duty by a simple warning notice, providing that in all the circumstances it is sufficient. Thus, in *Rae v Mars UK* (1989), it was held that where an unusual danger exists, the visitor should not only be warned of the danger but a barrier or additional notice should be placed to show its immediacy. As, however, it is possible for an occupier to discharge his duty by the simple expedient of a suitable notice, this supports our contention that the duty is not onerous. In addition, one should also note that there is no need for an occupier to warn of obvious dangers (*Staples v West Dorset District Council* (1995); *Darby v National Trust* (2001)), even where the visitor is a child (*Liddle v Yorkshire (North Riding) County Council* (1934)). If an occupier fails to warn of a particular danger, and the visitor suffers injury attributable to a different cause, no liability attaches to the occupier (*Darby*). This reinforces our conclusion that the duty is not onerous.

Finally, the 1957 Act allows the occupier to discharge his duty by entrusting work to independent contractors. Section 2(4)(b) provides that where damage is caused to a visitor by a danger due to the faulty execution of any work of construction, maintenance or repair by an independent contractor employed by the occupier, the occupier is not to be treated as answerable for the danger. This applies if, in all the circumstances, the occupier had acted reasonably in entrusting the work to an independent contractor, and had taken such steps (if any) as he reasonably ought in order to satisfy himself that the contractor was competent and that the work had been properly done. This sub-section has been liberally interpreted as regards the phrase 'work of construction, maintenance or repair'. In *AMF International v Magnet Bowling* (1968), it was held that the carrying out of some minor work was enough to bring the sub-section into operation and, in *Ferguson v Welsh* (1987), it was held that 'construction' included demolition. It was also held in *Ferguson* that although the sub-section requires the occupier to check that the work 'had been properly done', the sub-section could apply where the work was still in progress and had not been completed. These decisions may be seen as enlarging the circumstances in which an occupier owes a duty of care, but the courts are willing to find that entrusting work to an independent contractor is reasonable, and the wording of the sub-section contemplates that checking the work is not necessary in all circumstances. The guideline used by the courts is that the more technical the work, the less reasonable it is to require the occupier to check it. So, in *Haseldine v Daw* (1942), it was held that an occupier need not check the work of a firm of lift repairers, whereas in *Woodward v Mayor of Hastings* (1945), it was held that the work of a cleaner should be checked. Thus, although the courts have widened the scope of the sub-section as regards circumstances in which the duty will arise, the actual duty still seems to be similar to the familiar *Donoghue* duty to take reasonable steps. We could note here that the 1957 Act has made life less onerous for occupiers. Prior to the Act, it was held by the House of Lords in *Thomson v Cremin* (1956) that the duty an occupier owed to his visitors was a personal non-delegable duty that could not be discharged by employment of competent independent contractors.

We now need to consider the duty that an occupier owes to non-visitors or trespassers. A trespasser has been defined by Lord Dunedin in *Addie v Dumbreck* (1929) as a person who goes onto land without invitation and whose presence is either unknown to the occupier or, if known, is objected to.The duty owed to such a person is covered by the Occupiers' Liability Act 1984. By s 1(3) of this Act, an occupier will only owe a duty to a non-visitor if:

(a) he is aware of the danger or has reasonable grounds to believe that it exists;

(b) he knows or has reasonable grounds to believe that the non-visitor is in the vicinity of the danger or may come into the vicinity; and

(c) the risk is one against which, in all the circumstances, he may reasonably be expected to offer the non-visitor some protection.

In *Donoghue v Folkestone Properties Ltd* (2003), following *Ratcliff v McConnell* (1999) it was held that the duty under s 1(3) would not arise until the occupier had actual knowledge of the trespasser or of facts which made it likely that the trespasser will come onto the land. It was emphasised that actual knowledge was required but that the occupier was under no duty to the trespasser to ascertain whether or not such facts do exist. Once the occupier has actual knowledge of such facts, he will owe a duty if a reasonable occupier would realise the risk and the likelihood of the trespassers' presence. Finally the likelihood of the trespassers' presence at the actual time and place must be considered.

If the duty does arise, it consists of taking reasonable steps to enable the trespasser to avoid the damage.

Thus, in *Donoghue*, it was held that the defendant council could not reasonably expect a person to swim in the area in question in the middle of the night in mid-winter. It was also held in *Tomlinson v Congleton Borough Council* (2003) that the occupier owed no duty as regards dangers that were obvious. In both *Donoghue* and *Tomlinson*, the court drew a distinction between injuries caused by the state of the premises and injuries caused by the trespasser indulging in dangerous activities, the occupier owing no duty in respect of the latter. In *Tomlinson*, their Lordships stated that it was important that persons should accept responsibility for risks that they freely chose to take, and the activities of responsible persons should not be prohibited purely to safeguard the irresponsible minority from obvious risks. Their Lordships also stated that in determining what an occupier should do under the 1984 Act as well as the likelihood of injury and the probable severity of such injury, the cost of preventive measures and the social value of the activity giving rise to the risk should be considered.

The whole tenor of the judgment in *Tomlinson* means that the duty now owed to non-visitors is far from onerous.

Having considered the nature of the occupier's duty, we now need to see to what extent it can in fact be excluded. Section 2(1) allows an occupier to exclude his duty insofar as he is free to 'by agreement or otherwise'. A major restriction on the freedom of the occupier to do this is contained within the Unfair Contract Terms Act 1977. The 1977 Act controls, *inter alia*, the exclusion of liability for negligence and, by s 1(1)(c) of the 1977 Act, this includes the common duty of care imposed by the 1957 Act. Section 2(1) of the 1977 Act renders void any attempt to exclude liability for death or personal injury resulting from negligence. By s 2(2), attempts to exclude liability for other loss or damage are subject to the requirement of reasonableness. However, the 1977 Act only applies to business liability (s 1(3)), and so an occupier of (say) a private house may exclude the duty he would otherwise owe to his visitors by a suitable exclusion clause.[1]

An occupier may be exonerated from liability for risks willingly accepted as his by the visitor (s 2(2); *Simms v Leigh Rugby Football Club* (1969)) and, in deciding whether a visitor is *volens* ('consented' to) to a danger, the presence of warnings or exclusion notices may be relevant. Similarly, the liability of an occupier may be reduced by contributory negligence on the part of the visitor (*Stone v Taffe* (1974)) and again the presence of such notices may be relevant.

Overall, therefore, it is difficult to agree with the statement that the duty an occupier owes to his visitors or non-visitors is onerous. It is based on the requirement to act reasonably and can be discharged relatively easily through warning notices or the use of independent contractors. As the duty is to take reasonable steps, it could be argued that the onerousness or otherwise of the duty will depend on what the courts consider to be reasonable conduct on the part of an occupier. In this context, we might note the *dicta* of the House of Lords in *Smith v Littlewoods Organisation Ltd* (1987), where both Lords Goff and Mackay were at pains to emphasise that no unreasonable burdens should be placed on occupiers, so it would seem unlikely that any more would be required of occupiers by the courts in the future than has been done in the past. In addition, a non-business occupier will be free to exclude his duty, whilst a business occupier will be free to exclude his liability for loss or damage other than death or personal injury insofar as his exclusion term satisfies the requirements of reasonableness.

Think point

1 There is another possible restriction on the freedom of an occupier to exclude his duty. It has been argued in Rogers (ed), *Winfield and Jolowicz on Tort*, 16th edn, 2002, that the standard of care imposed by the House of Lords in *British Railways Board v Herrington* (1972), which applied to trespassers, represents a minimum standard that cannot be excluded, as it was based on a standard of common humanity. However, there is no authority either for or against this proposition.

Notes

8 Nuisance

Introduction

Questions on nuisance are popular with examiners, possibly because nuisance is a complex topic with several unresolved areas. Much of this complexity is due to the fact that there are few hard and fast rules as to what constitutes a nuisance; instead, there are a number of guidelines which the court may or may not decide are relevant in deciding whether a particular activity amounts to a nuisance.

Recently, a number of nuisance cases have involved the provisions of the Human Rights Act 1998, and candidates should be aware of the importance of this rapidly developing area of law.

Checklist

Students must be familiar with the following areas:

(a) types of activity capable of constituting a nuisance;

(b) factors indicating whether an interference is unreasonable and the relative importance of these factors *inter se*;

(c) possible defendants in a nuisance action;

(d) defences and especially invalid defences;

(e) the undecided point regarding recoverability of damage for personal injury and economic loss;

(f) public nuisance; and

(g) the relevance of the Human Rights Act 1998, especially s 6, and Arts 2 and 8 of the European Convention on Human Rights.

In addition, a nuisance question may contain elements of negligence or *Rylands v Fletcher* (1868).

Question 25

Sarah owns a house in a small village which she leases to May. May owns four dogs which she keeps in kennels in the garden. The dogs spend large amounts of the day and night barking and this annoys her neighbours, Terence and Ursula. Victor, another neighbour, finds the noise during the day particularly annoying, as he works nights and has to sleep during the day. All the neighbours complain to May, who refuses to do anything. Consequently, Ursula lights a large bonfire in her garden in the hope that the smoke will stop the barking. Terence, whose hobby is woodworking, takes the television suppressor off his electric drill and uses it in the evenings to interfere deliberately with the reception on May's television.

Discuss the legal situation.

Answer plan

The following points need to be discussed:

- whether the barking of the dogs is a nuisance;
- liability of landlord and tenant in nuisance;
- whether Victor is a sensitive claimant;
- liability of Ursula in nuisance for the bonfire;
- liability of Terence in nuisance for the interference with the TV reception; and
- liability of the local authority under the Human Rights Act 1998.

Answer

We must first decide whether the barking of the dogs constitutes a nuisance. A nuisance is an unreasonable interference with a person's use or enjoyment of land or some right over or in connection with it. It is well-established that noise can constitute a nuisance (*Halsey v Esso Petroleum* (1961); *Tetley v Chitty* (1986)), but not all interference gives rise to liability. There must be give and take between neighbours and the interference must be substantial and not fanciful (*Walter v Selfe* (1851)). As we are told that the dogs spend large amounts of the day and night barking, this noise would amount to a nuisance. Duration of the interference is one of the factors that a court would take into account in deciding whether a noise amounts to a nuisance. The shorter the duration of the interference, the less likely it is to be unreasonable (*Harrison v Southwark and Vauxhall Water Co* (1891)).

Given that the barking of the dogs constitutes a nuisance, we must next decide who is the proper defendant in respect of this nuisance. As May is responsible for the dogs, she will be a defendant. The landlord, Sarah, will not be liable: the nuisance did not exist before she leased the premises; the premises have not been let for a purpose which constitutes a nuisance (as in *Tetley*); and any right Sarah has reserved to enter and repair is irrelevant, as the nuisance has not arisen due to the disrepair of the premises. Hence, May is the only defendant.

Finally, we must ascertain who can sue in respect of the barking. As nuisance is concerned with a person's use or enjoyment of land, it was held that only persons with an interest in land can sue (*Malone v Laskey* (1907)). On this basis, if Terence and Ursula are owners or tenants of the property, they can sue but if, for example, Terence is the sole owner, Ursula would not have the requisite interest in land traditionally required to found an action in private nuisance. Although in *Khorasandjian v Bush* (1993) and *Hunter v Canary Wharf Ltd* (1996), the Court of Appeal that it was no longer necessary to have the classic interest in land required by *Malone*, when *Hunter* was decided in the House of

Lords, the House overruled the Court of Appeal on this point and held that *Malone* was still good law. Victor can sue if he has the traditional interest in land, but may run into the problem that being a night worker, he is a sensitive claimant. In *Robinson v Kilvert* (1889), it was held that a claimant cannot recover where the damage is solely due to the sensitive nature of the claimant's property. However, in *McKinnon Industries v Walker* (1951), it was held that once a nuisance has been established on the grounds of interference with ordinary use, a claimant can recover for interference with a sensitive use. Hence, if Victor can establish that the barking constitutes an unreasonable interference with his use or enjoyment of property, he will have full remedies. Any harm caused must be foreseeable (*The Wagon Mound (No 2)* (1967); *Cambridge Water Co v Eastern Counties Leather plc* (1994)), but this requirement gives rise to no problems for the claimants.

The remedies available against May would be damages to compensate for past nuisance and an injunction to prevent further nuisance. The court does have power, under s 50 of the Supreme Court Act 1981, to award damages in lieu of an injunction, but this power is used very sparingly. In *Shelfer v City of London Electric Lighting Co* (1895), the Court of Appeal held that damages should only be awarded where:

(a) the injury to the claimant's legal rights is small;

(b) the damage is capable of being estimated in money;

(c) the damage can be adequately compensated by a small money payment; and

(d) the case is one in which it would be oppressive to the defendant to grant an injunction.

In *Jaggard v Sawyer* (1995), the Court of Appeal stated that these criteria provided a good working rule, but that the basic question was whether, in all the circumstances, it would be oppressive to the defendant to grant the injunction. On the facts that we are given, there seems no good reason for the court to diverge from the normal practice, and thus an injunction should granted against May.

We must next consider the actions of Ursula and Terence. Ursula has lit a large bonfire in her garden. This of itself may not constitute a nuisance – the interference must be substantial and not merely fanciful (*Walter v Selfe* (1851)). In deciding whether a particular interference is unreasonable or not, the court will rely on a series of guidelines, rather than on any rigid rules. In Ursula's case, the court would consider the duration of the interference, as the shorter the duration of the interference, the less likely it is to be unreasonable (*Harrison v Southwark and Vauxhall Water Co* (1891)). In particular, it seems that an isolated event is unlikely to constitute a nuisance. In *Bolton v Stone* (1951), it was stated that a nuisance must be a state of affairs, however temporary, and not merely an isolated happening. Thus, although Ursula might claim that the bonfire is an isolated event, it does constitute a temporary state of affairs and is capable, in law, of being a nuisance. A possible argument that Ursula might employ is that she only lights a bonfire on rare occasions and that this is a reasonable use of her land. However, the fact that a defendant is only making reasonable use of his land is not, of itself, a valid defence in nuisance (*AG v Cole* (1901); *Vanderpant v Mayfair Hotel* (1930)). As regards any interference with health and comfort, the court will take into account the character of the neighbourhood (*Bamford v Turnley* (1860)), as 'what would be a nuisance in Belgravia Square would not necessarily be so in Bermondsey' (*Sturges v Bridgman* (1879), *per* Thesiger LJ). Thus, as Ursula lives in a rural area, the occasional lighting of a bonfire might not constitute a nuisance, as there must be an element of give and take between neighbours. However, if Ursula by her lack of care allowed an annoyance from the bonfire to become excessive, she would become liable in nuisance (*Andreae v Selfridge* (1938)). The character of the neighbourhood is not relevant, however, if the nuisance causes physical damage to the property (for example, if smoke from the bonfire discolours paintwork, etc) (*St Helen's Smelting Co v Tipping* (1865)). The real problem that Ursula faces, however, is that she is activated by malice. Although malice is not a necessary ingredient of nuisance, its presence is not only a factor to be taken into account (*Christie v Davey* (1893)), but may even turn an otherwise non-actionable activity into a nuisance (*Hollywood Silver Fox Farm v Emmett* (1936)), where it seems clear that in the absence of malice no action would have arisen. Ursula cannot claim to be making a reasonable use of her land if she intends to cause damage. Hence, May could sue Ursula in nuisance (May having the necessary interest in land, being a tenant) and obtain damages and an injunction.

Terence is deliberately interfering with May's television reception. This interference is presumably not of limited duration, which is a factor mitigating against unreasonableness of any interference (*Harrison v Southwark and Vauxhall Water Co* (1981)). Terence is clearly activated by malice, as in *Hollywood Silver Fox Farm* and *Christie*, so *prima facie* he would seem to have committed a nuisance. However, May has a problem, in that, in *Bridlington Relay v Yorkshire Electricity Board* (1965), Buckley J held that interference with purely recreational facilities, such as television reception, did not constitute an actionable nuisance and stated that 'at present' the ability to receive interference-free television signals was not so important a part of a householder's enjoyment of his property as to be protected in nuisance. The phrase 'at present' has been quoted by later claimants and, in the Canadian case of *Nor-Video Services Ltd v Ontario Hydro* (1978), it was held that interference with television reception could amount to a nuisance, and the High Court in *Hunter* took a similar view. Unfortunately for May, both the Court of Appeal and the House of Lords in *Hunter* rejected this approach, holding that interference with television reception did not constitute an actionable nuisance. Thus, May has no remedy in respect of Terence's actions.

Finally, we should consider any remedies that might be available to Terence, Ursula and Victor under the Human Rights Act 1998. Under Art 8 of the European Convention on Human Rights, brought into UK law by s 1 of the 1998 Act, Terence, Ursula and Victor have the right to respect for private and family life. In *López Ostra v Spain* (1995), it was held that the construction of a waste treatment plant next to the applicant's house which caused local pollution and health problems was a violation of the applicant's rights. In this case, the Spanish Government did not own the plant, but it was sufficient that the local authority had allowed the plant to be built. Under s 6(1) of the 1998 Act, it is unlawful for a public authority to act in contravention of a Convention right. Also, and by s 6(6), 'act' includes a failure to act. Consequently, it could be argued that by failing to institute proceedings for statutory nuisance, the local authority has failed to protect the rights Terence, Ursula and Victor have under s 6 and are thus liable in damages to these parties. In *Baggs v UK* (1987) and *Hatton v UK* (2003), it was taken for granted by the European Court of Human Rights that noise came within Art 8.

Notes

Question 26

Beta Products plc owns a factory set in the centre of a manufacturing town and employs a considerable number of people. One day, the factory emits a quantity of acid fumes which damage the paintwork of the neighbouring houses and some residents' cars. In addition, Beta has recently installed some machinery which is considerably more noisy than the machinery it replaced, and which annoys their immediate neighbours.

Discuss any potential liability of Beta both at common law and under the Human Rights Act 1998.

Answer plan

This question seems only to cover a few issues of nuisance. However, careful study will show that it raises a number of common law issues, as well as the rapidly developing area of Human Rights Act actions.

The following points need to be discussed:

- utility of Beta's conduct;
- relevance of neighbourhood to interference with health and comfort and physical damage;
- whether an isolated event can constitute a nuisance;
- possibility of action in public nuisance;
- additional actions in *Rylands v Fletcher* (1868) and negligence; and
- situation under the Human Rights Act 1998, with especial reference to Art 8 of the European Convention on Human Rights.

Answer

We shall first consider whether Beta plc has incurred any liability in nuisance. A nuisance is an unreasonable interference with a person's use or enjoyment of land, or of some right over or in connection with it. However, not all interference necessarily gives rise to liability and there must be give and take between neighbours. Also, the interference must be substantial and not fanciful (*Walter v Selfe* (1851)). The courts have developed a number of guidelines that are used to determine whether any particular interference is unreasonable, but each test is only a guideline and not a condition, and the court has to evaluate the defendant's behaviour in all the circumstances of the case. In Beta's case, the court will consider whether the emission was an isolated event. In *Bolton v Stone* (1951), it was stated that a nuisance could not arise from an isolated happening, but had to arise from a state of affairs, however temporary. Thus, in *Midwood v Manchester Corp* (1905), a gas explosion was held to be a nuisance because, although it was an isolated event, it was due to a pre-existing state of affairs, namely, the build up of gas. On this basis, it could be argued that the escape of acid smut was due to a build up of this material on Beta's premises and thus the emission can constitute an actionable nuisance. The damage suffered is not due to any sensitive use of the neighbours' property (as in *Robinson v Kilvert* (1889)) and, although the premises are in the centre of a manufacturing town, the character of the neighbourhood is not to be taken into account where physical damage to property has been caused (*St Helen's Smelting Co v Tipping* (1865)). We are told that Beta employs a considerable number of people, but the utility of the defendants' conduct, although a factor to be taken into account, seems to be a factor of lesser importance in the overall assessment (*Adams v Ursell* (1913); Irish case of *Bellew v Cement Co* (1948)). Thus, the fact that Beta provides employment is not a conclusive factor. Note that it would not be necessary to show that Beta was negligent, as negligence is not an essential ingredient of nuisance. Indeed, it would be no defence to Beta to show that it took all reasonable care and even all possible care; provided that it caused the nuisance, that is sufficient. Thus, taking all the circumstances into account, a court would find that the emission constituted an actionable nuisance. Traditionally, an interest in the land in question was required as a prerequisite for an action in nuisance (*Malone v Laskey* (1907)) and, although this requirement was relaxed by the Court of Appeal in *Khorasandjian v Bush* (1993) and *Hunter v Canary Wharf Ltd* (1996), it was reimposed by the House of Lords in *Hunter* (1997). Hence, only owners and tenants of the properties affected may sue, and not their guests or lodgers. The owners and tenants could obtain damages for the loss (apart from any personal injuries) they have suffered, together with an injunction to prevent future emissions.

As regards the noisy machinery, it is well-established that noise can constitute a nuisance (*Halsey v Esso Petroleum* (1961); *Tetley v Chitty* (1986)). In deciding whether the noise from the machinery amounts to a nuisance, it is clearly not an isolated event of limited duration, nor is there any evidence of sensitivity on the part of the neighbours. However, as the noise is an interference with health and comfort, the character of the neighbourhood must be taken into account (*Bamford v Turnley* (1860)). As Thesiger LJ stated in *Sturges v Bridgman* (1879), 'what would be a nuisance in Belgravia Square would not necessarily be so in Bermondsey'. As we are told that Beta's factory is in the centre of a manufacturing town, the neighbours would have to accept a certain amount of noise as part of everyday living. However, in *Roshner v Polsue and Alfieri Ltd* (1906), where a person lived in an area devoted to printing, he obtained an injunction to prevent the use of a new printing machine which interfered with his sleep. Thus, it will be a question of fact for the court to decide whether or not the increased noise amounts to a nuisance in all the circumstances of the case. Again, only persons with an interest in land could sue in respect of this noise (*Malone*; *Hunter*). It should also be noted that foreseeability of harm is a necessary ingredient of nuisance (*The Wagon Mound (No 2)* (1967); *Cambridge Water Co v Eastern Counties Leather plc* (1994)), but harm is foreseeable for both the fumes and the noise.

Beta might also, as regards the emission, be liable in public nuisance (*Halsey v Esso Petroleum* (1961)). Similar considerations will apply as for private nuisance, but some additional factors must be shown. First, the persons affected by the nuisance must consist of the public or a section of the public (*AG v PYA Quarries* (1957)). Secondly, the claimant must have suffered damage over and above that suffered by the public at large. In *Halsey,* it was held that where acid smuts damaged washing hung out to dry and a car, the owner of the damaged property could sue in public nuisance. Thus, the car owners whose car paintwork is damaged could sue in public nuisance and they would not have to have any interest in land. Whether those persons whose paintwork was damaged could sue would depend on their being able to prove damage over and above that suffered by the public at large.

An additional cause of action which might lie against Beta is under the rule in *Rylands v Fletcher* (1868). Thus, a person who for his own purposes brings onto his land and collects and keeps there anything likely to do mischief if it escapes must keep it in at his peril and, if he does not do so, he is *prima facie* answerable for all the damage which is the natural consequence of its escape. In addition, there must be a non-natural user of land and the damage must be foreseeable (*Cambridge Water Co*). Again, in *Halsey*, the defendants were liable under *Rylands* for the damage caused by the acid smuts to both the washing and the car. A problem that could arise is whether Beta has made a non-natural use of its land. After *Halsey*, it was held in *British Celanese v Hunt* (1969) that factories in industrial parks were a natural use of land. More recently however, in *Cambridge Water Co*, the House of Lords took a more restrictive approach as to what constitutes natural use, and in particular stated that the provision of employment did not, of itself, constitute a natural or ordinary use of land. In *Transco v Stockport Metropolitan Borough Council* (2004), the House of Lords undertook an extensive, in-depth review of the rule in *Rylands* and held that the requirement that the thing is likely to do mischief if it escapes should not be easily satisfied. It must be shown that the defendant has done something which he recognised or ought to have recognised as giving rise to an exceptionally high risk of danger or mischief if it escapes, however unlikely such an escape might have been thought to be. The acid fumes and noise come into this category. In addition, their Lordships considered the non-natural use requirement and held that the defendant's use of the land must be extraordinary and unusual. The House doubted whether the test of reasonable use was helpful, since a use may be very out of the ordinary but still reasonable, such as the storage of chemicals in an industrial park. Thus, the fact that Beta's factory is situated in a manufacturing town, which might be relevant to reasonable use, still does not stop the use from being extraordinary and unusual. Hence it seems likely that Beta could be liable in a *Rylands* action.

It would also be possible for those persons affected by the emission to sue in negligence. The claimants will have to show that Beta owes them a duty of care. In a novel fact situation, the court will apply the test favoured by the House of Lords in *Caparo Industries plc v Dickman* (1990) and *Murphy v Brentwood District Council* (1990), namely, to consider the foreseeability of damage, proximity of relationship and the reasonableness or otherwise of imposing a duty of care. If a duty is found to exist, it must be shown that Beta was in breach of that duty by failing to act as a reasonable factory owner would (*Blyth v Birmingham Waterworks* (1856)). It must also be shown that this breach caused the damage, and that the damage was not too remote, in that it was reasonably foreseeable (*The Wagon Mound (No 1)* (1961)). The problem in a negligence action will be in proving that Beta was in breach of their duty for, if they followed the standard procedure of their trade, that is good evidence they were not in breach (see, for example, *Knight v Home Office* (1990)).

Finally, we must consider any causes of action that might arise under the Human Rights Act 1998. Article 8 of the European Convention on Human Rights, which was brought into UK law by s 1 of the Human Rights Act 1998, establishes the right to respect for private and family life and home. Also, Art 1 of the First Protocol states that persons are entitled to the peaceful enjoyment of their possessions and Art 2 establishes a right to life.

As regards Art 8, in *López Ostra v Spain* (1995), it was held that the construction of a waste treatment plant next to the applicant's house, which had caused local pollution and health problems, was a violation of Art 8. In this case, the Spanish Government did not own the plant, but it was held to be sufficient that the local authority had allowed it to be built on their land and the Government had subsidised it. As s 6(1) of the 1998 Act makes it unlawful for a public authority to act in any way incompatible with a Convention right, and by s 6(6) an 'act' includes a failure to act, both the Government and local authorities could be held liable for breaches of Art 8. Article 8 has also been held to apply to toxic emissions from a factory (*Guerra v Italy* (1998)), so clearly hazardous emissions could fall within Art 8 and even Art 2 if the emissions were sufficiently hazardous. However, in *Hatton v UK* (2003) and *Marcic v Thames Water Utilities Ltd* (2004), the fundamentally subsidiary nature of the Convention was emphasised. In *Hatton,* it was stated that national authorities have direct democratic legitimation and are well-placed to evaluate local needs and conditions. Thus, if the local authority had canvassed opinions of persons affected by the activities of Beta and considered these opinions before allowing the activities, it is possible that the Convention action would fail: see *Hatton.*

An action under the Human Rights Act 1998 would raise no problems as regards interest in land (*McKenna v British Aluminium Ltd* (2002)), recovery of economic loss or application to personal injuries. Indeed, in *Marcic v Thames Water Utilities Ltd* (2001), the High Court judge found for the claimant under Art 8, while dismissing the claims based on nuisance and *Rylands v Fletcher*; although the House of Lords dismissed the claimants' actions in both nuisance and under Art 8 (following *Hatton*), it is clear from the High Court decision that much of the detailed law of nuisance is irrelevant in considering a breach of Art 8. Thus, an action under the Human Rights Act 1998 would be available to a considerable range of claimants who suffer personal injury due to Beta's factory (subject to the subsidiary nature point discussed in *Hatton* and *Marcic*), and the possible defendants to such an action could be the local authority under s 6(1) and (6) of the 1998 Act, or the UK Government.

Notes

Question 27

The Northwood Council has run an adventure centre for young adolescents for many years. The centre includes a go-kart racetrack. About a year ago, several houses were built adjacent to the centre and the residents of these houses now complain of the noise from the racetrack. One resident, George, claims that the noise has further impaired his hearing, which was already damaged due to his having worked in a noisy environment for many years.

Advise the residents, including George, of any remedies available to them.

Answer plan

This is a deceptively simple question that covers a range of aspects of both liability and defences to actions in nuisance, together with the possibility that must always be considered in nuisance cases of alternative courses of action.

The following points need to be discussed:

- whether the noise constitutes a nuisance;
- liability of the Council for any nuisance, private or public;
- possible defences available to the Council;
- action under *Rylands v Fletcher* (1868);
- action in negligence; and
- liability of the Council under the Human Rights Act 1998.

Answer

Dealing first with the noise emanating from the centre, it is well-established that noise is capable of constituting a nuisance (*Halsey v Esso Petroleum* (1961); *Tetley v Chitty* (1986)). A nuisance can be defined as an unreasonable interference with a person's use or enjoyment of land, or some right over or in connection with it. However, not all interference will necessarily constitute a nuisance: there must be give and take between neighbours and the interference must be substantial and not fanciful (*Walter v Selfe* (1851)). There are a number of factors that the court takes into account in deciding whether an interference is unreasonable or not, and we shall consider the application of these guidelines to the noise in question. One factor that needs to be considered is the duration of the interference, since if this is short, the interference is not likely to be held unreasonable (*Harrison v Southwark and Vauxhall Water Co* (1891)). However, we are told that the activity centre has been in operation for many years, so this time factor is in favour of the existence of a nuisance. We are not told that the claimants are especially sensitive to noise or that a reasonable person living in the area would not object to the noise, which are capable of being possible defences, so we must consider the character of the neighbourhood. This is a relevant factor where the interference is with health and comfort (*Bamford v Turnley* (1860)), as Thesiger LJ stated in *Sturges v Bridgman* (1879), 'What would be a nuisance in Belgravia Square would not necessarily be so in Bermondsey'. In *Halsey v Esso Petroleum* (1961), Veale J held that the standard was that of the ordinary and reasonable man living in the vicinity of the alleged nuisance. This would be a question of fact for the court to decide. The fact that the centre is socially useful is a factor to be considered, but it seems to be easily overridden (see *Adams v Ursell* (1913); *Bellew v Cement Co* (1948)).

Given that the noise can be shown to be unreasonable, the next question is who can sue. Formerly, an interest in the land affected was required as a prerequisite for suing in nuisance (*Malone v Laskey* (1907)) and, although this requirement was relaxed by the Court of Appeal in *Khorasandjian v Bush* (1993) and *Hunter v Canary Wharf Ltd* (1996), it was reimposed by the House of Lords in *Hunter* (1997). Hence, only owners and tenants of the properties affected may sue, and not their guests or lodgers. It should also be noted that foreseeability of harm is a necessary ingredient of nuisance (*The Wagon Mound (No 2)* (1967); *Cambridge Water Co v Eastern Leather plc* (1994)), but there is no difficulty in showing foreseeability of harm in respect of the noise.

We must next decide on an appropriate defendant. Clearly, there would be little point in suing the adolescents and so the only defendant for practical purposes would be the Council. The owner of land may be liable for a nuisance committed on his land which he has not created where he allows the land to be used for a purpose, and a nuisance is an 'ordinary and necessary' consequence of such use (*Tetley v Chitty* (1986), *per* McNeill J). Similarly, a landowner may also be liable where he allows persons to use a lane as a base from which to disturb the claimants (*Lippiatt v South Gloucestershire Council* (1999)). Thus, the neighbours could sue the Council for damages in respect of past noise nuisance and for an injunction to stop future noise nuisance. It would be no defence to the Council to allege that the claimants came to the nuisance (*Sturges v Bridgman* (1879)), although the claimants must of course accept the standard of the neighbourhood to which they come. Neither would it be possible for the Council to claim the defence of prescription, that is, that the nuisance has been continued for 20 years, because time does not begin to run until claimants are aware of the nuisance (*Sturges*). Nor would it avail the Council to claim that the public interest of providing an adventure centre should prevail over the private rights of residents (*Pride of Derby v British Celanese* (1953); *Kennaway v Thompson* (1981), where private interests were held to prevail over public interests).[1]

From the facts given, it seems unlikely that a sufficient number of persons are affected for the activities to amount to a public nuisance (*AG v PYA Quarries* (1957)).

As regards George, he has the problem that he may be an abnormally sensitive claimant. The standard, as stated previously, is that of the ordinary and reasonable man living in the vicinity of the alleged nuisance and, where the damage is entirely to an abnormal sensitivity on the part of the claimant, no action will lie (*Robinson v Kilvert* (1889); *Heath v Mayor of Brighton* (1908)). Thus, on this ground, George could not recover damages in respect of his additional hearing loss. If, however, it can be shown that a nuisance does exist, George will not be denied damages or an injunction to stop the nuisance merely because of his sensitivity (*McKinnon Industries v Walker* (1951)). George has an additional problem in recovering for his fresh impaired hearing loss, in that it was an undecided point whether damages in respect of personal injury can be recovered in nuisance. In *Cunard v Antifyre* (1933), it was stated that recovery for personal injury was not possible. Also, in *Cambridge Water Co v Eastern Counties Leather* (1994), the House of Lords, when considering an action under the rule in *Rylands v Fletcher* (1868), referred with approval to a 'seminal' article by Professor Newark ('The boundaries of nuisance' (1949) 65 LQR 480), in which he argued that recovery for personal injury should not be possible in nuisance. Although the House did not decide the situation regarding recovery in respect of personal injury in *Rylands*, let alone nuisance, their Lordships' wholehearted acceptance of Professor Newark's article suggests that a future court would not allow such recovery. Thus, it came as no surprise when, in *Hunter v Canary Wharf Ltd* (1997), which was an action in nuisance, the House of Lords followed their reasoning in *Cambridge Water Co* and stated that actions for personal injury should not be brought in nuisance. Hence, George cannot recover in nuisance in respect of his additional hearing loss.

The residents could also sue the Council in negligence. The residents will have to show that the Council owes them a duty of care. In a novel fact situation, the court will apply the test favoured by the House of Lords in *Caparo Industries plc v Dickman* (1990) and *Murphy v Brentwood District Council* (1990), namely, to consider the foreseeability of damage, proximity of relationship and the reasonableness or otherwise of imposing a duty of care. If a duty is found to exist, it must also be shown that the Council was in breach of its duty by failing to act as a reasonable Council would (*Blyth v Birmingham Waterworks* (1856)); that this breach caused the damage (*Cork v Kirby MacLean* (1952)); and that the damage was not too remote, in that it was reasonably foreseeable (*The Wagon Mound (No 1)* (1961)). On the facts, it is submitted that the residents could succeed in establishing the elements of negligence on the part of the Council. However, the Council might run the defence of lack of funds to carry out any suitable sound reducing measures, such as landscaping or erection of sound absorbing barriers. In *Knight v Home Office* (1990), in holding that a prison hospital had not been in breach of duty, Pill J stated that the court must take into account the fact that resources available for the public sector are limited. However, as the cricket club in *Miller v Jackson* (1977) was held liable in negligence, it would seem likely that the Council would also be found liable.

George would be in a much stronger position in negligence, as he could recover for personal injury, and the fact that he is abnormally sensitive as regards excessive noise is immaterial as, in negligence, the defendant takes his claimant as he finds him, providing that some foreseeable damage occurs (*Dulieu v White* (1901); *Smith v Leech Brain* (1962)).

The residents should also be advised of the chances of mounting a successful action under the rule in *Rylands*. The rule states that a person who, for his own purposes, brings onto his land and collects there anything likely to do mischief if it escapes must keep it in at his peril and, if he does not do so, he is *prima facie* answerable for all the damage which is the natural consequence of its escape. In addition, the defendant must make a non-natural use of his land. Although the noise and the go-karts have not been collected and kept by the Council on their land, it could be argued that the Council have incurred liability by allowing the presence of the go-karts and the escape of the noise. This argument is strengthened by the acceptance by the House of Lords in *Cambridge Water Co* of the view that *Rylands* is an example of nuisance applied to an isolated escape and, as we have seen, the Council would be liable in nuisance. The next problem is whether the Council has made a non-natural use of its land. In *Transco v Stockport Metropolitan Borough Council* (2004), the House of Lords undertook an extensive, in-depth review of the rule in *Rylands* and held that the requirement that the thing is likely to do mischief if it escapes should not be easily satisfied. It must be shown that the defendant has done something which he recognised or ought to have recognised as giving rise to an exceptionally high risk of danger or mischief if it escapes, however unlikely such an escape might have been thought to be. Noise can come into this category. In addition, their Lordships considered the non-usual use requirement and held that the defendant's use of the land must be extraordinary and unusual. The House doubted whether the test of reasonable use was helpful, since a use may be very out of the ordinary but still reasonable, such as the storage of chemicals in an industrial park. Thus, the fact that the adventure centre may have a social value might be relevant to reasonable use, but does not stop the use from being extraordinary and unusual. Thus, as the harm is foreseeable (as in nuisance), the Council may be liable under the rule in *Rylands*.

Finally, we should consider any remedies that might be available to the residents under the Human Rights Act 1998. Under Art 8 of the European Convention on Human Rights, brought into UK law by s 1 of the 1998 Act, the residents have the right to respect for their private and family life. In *López Ostra v Spain* (1995), it was held that the construction of a waste treatment plant next to the applicant's house, which caused local pollution and health problems, was a violation of the applicant's rights. In this case, the Spanish Government did not own the plant, but it was sufficient that the local authority had allowed the plant to be built. Under s 6(1) of the 1998 Act, it is unlawful for a public authority to act in contravention of a Convention right, and by running the go-kart racetrack it could be claimed that Northwood Council is acting in a way that is incompatible with the residents' Convention rights. In *Baggs v UK* (1987) and *Hatton v UK* (2003), it was taken for granted by the European Court of Human Rights that noise came within Art 8. This route would have the advantage that the residents would not have to show an interest in land to proceed under Art 8. However, in *Hatton v UK* (2003) and *Marcic v Thames Water Utilities Ltd* (2004), the fundamentally subsidiary nature of the Convention was emphasised. In *Hatton*, it was stated that national authorities have direct democratic legitimation and are well-placed to evaluate local needs and conditions. Thus, if the Council had canvassed opinions of the persons affected by the activities of the racetrack and considered these opinions before allowing the activities, it is possible that the Convention action would fail: see *Hatton*.

Think point

1 One could mention here the views expressed by Lord Denning in *Miller v Jackson* that public interests should prevail over private interests where there is a clash. However, this approach was not followed in *Kennaway*, where private rights were allowed to prevail. Lord Denning also stated that the reason for cricket balls coming into the plaintiff's garden was not the playing of cricket, but the building of the houses, but this view was rejected by the majority of the Court of Appeal. It seems unsafe to advise the residents in the present case to rely on this dissenting judgment.

   In addition, a landlord may be liable for the activities of his licencees if he allows the licencees to occupy his land and use it to disturb others, or may be deemed to have adopted the nuisance by failing to eject the licencees (*Lippiatt v South Gloucestershire Council* (1999)).

Notes

Question 28

'The difficulties of proceeding with an action in private nuisance are grave, but the prospects of potential claimants have increased with the coming into force of the Human Rights Act 1998.'

Discuss the above statement.

Answer plan

This question calls for a discussion of some of the problems that would be encountered in successfully running an action in private nuisance.

In particular, the following points need to be discussed:

- guidelines in determining whether any particular interference is unreasonable;
- possible defendants;
- defences available to a defendant;
- scope of the action as regards personal injury and economic loss; and
- nuisance and the Human Rights Act 1998.

Answer

The law on private nuisance, or nuisance as we shall henceforth call it, gives rise to a number of difficulties in its application to factual situations. This is due not to any conceptual difficulty, but rather to the variety of circumstances in which nuisances have been held to exist and to the flexible approach which the courts adopt in deciding in any given case whether or not a nuisance exists. In addition, the exact scope of the tort is shrouded in uncertainty. Therefore, we shall examine these uncertainties. A nuisance can be defined as an unreasonable interference with a person's use or enjoyment of land, or some right over or in connection with it. It was previously held from this definition that only persons with an interest in the land affected can sue (*Malone v Laskey* (1907)) and, although this requirement was relaxed by the Court of Appeal in *Khorasandjian v Bush* (1993) and *Hunter v Canary Wharf Ltd* (1996), it was reimposed by the House of Lords in *Hunter* (1997). Although this has narrowed the range of potential claimants, it has reintroduced some certainty back into nuisance, as the exact link between the person affected and the land was somewhat uncertain following the Court of Appeal decision and *dicta* in *Hunter*. It should also be noted that foreseeability of damage is a necessary ingredient of nuisance (*The Wagon Mound (No 2)* (1967); *Cambridge Water Co v Eastern Counties Leather plc* (1994)), although this is unlikely to be a problem in practice. It also seems, from the decision in *Bridlington Relay Ltd v Yorkshire Electricity Board* (1965), that interference with purely recreational facilities lies outside the tort of nuisance. In *Bridlington*, the court was concerned with the reception of interference-free television signals. Despite the use of the phrase 'at present' by Buckley J in his judgment, and the willingness of the High Court in *Hunter* (1994) to allow an action in nuisance for interference with television signals both the Court of Appeal (1996) and the House of Lords (1997) in *Hunter* held that an action did not lie for such interference. Thus, an uncertainty has been removed, in that the possibility of an action in such circumstances certainly does not arise, although the exact extent of purely recreational facilities is not clear.

Moving on to what constitutes an unreasonable interference, we meet a major area of uncertainty. The courts have laid down a series of guidelines as to what constitutes an unreasonable interference but, as in any situation where it has to be decided whether or not some particular conduct is reasonable, the courts' decisions cannot amount to binding precedents. The total circumstances of the case must always be taken into account in deciding this question. What gives rise to particular uncertainty in nuisance is that the courts seem willing, when the circumstances require it, to either disregard a particular guideline or to assign it less importance in some cases than in others. Nevertheless, there is one guideline that the courts seem willing to follow on almost all occasions, namely, the rule that not all interference gives rise to liability, that there must be give and take between neighbours and that the interference must be substantial and not merely fanciful (*Walter v Selfe* (1851)). When we consider the guidelines that the courts adopt, we shall see that there are three that the courts tend to apply in the majority of cases, and three that the courts consider, but which they seem more willing to attach a lower importance to if the circumstances so require.

Turning now to the first category of criteria, we have the duration of the interference. The shorter the duration of the interference, the less likely it is to be found unreasonable. So, in *Harrison v Southwark and Vauxhall Water Co* (1891), temporary work in sinking a shaft was held not to constitute a nuisance because of the temporary nature of the work. Given that a short interference is not likely to give rise to liability, the question arises as to whether an isolated event is capable of constituting a nuisance. In *Bolton v Stone* (1951), it was held that an isolated happening could not constitute a nuisance, but that what was required was a state of affairs, however temporary. Thus, in *Midwood v Manchester Corp* (1905), a gas explosion was held to constitute a nuisance, even though it was an isolated event, because it was due to a pre-existing state of affairs, namely, a build-up of gas. *Castle v St Augustine's Links* (1922) is a similar example of an isolated event being held to constitute a nuisance, as the occurrence was due to a pre-existing wrongful state of affairs. One factor which the courts seem to always take into account is whether the claimant is abnormally sensitive. The rule is that a person cannot increase his neighbour's possible legal liability just because he puts his land to some special use. Thus, in *Robinson v Kilvert* (1889), a plaintiff could not recover for damage caused by heat from the defendant's heating pipes to his stock of 'exceptionally sensitive' brown paper, as the heat would not have interfered with a normal use of the property. However, once a nuisance has been established, full remedies are available in respect of any unusually sensitive use the claimant makes of his property (*McKinnon Industries v Walker* (1951)). The character of the neighbourhood is also a relevant factor where the interference is with health and comfort (*Bamford v Turnley* (1860)). This is also illustrated by the famous statement of Thesiger J in *Sturges v Bridgman* (1879), where he said 'what would be a nuisance in Belgravia Square would not necessarily be so in Bermondsey'. It should not be thought that this criterion means that if an area is industrialised or built up, then no nuisance can take place there: the question that has to be decided is whether the interference is unreasonable or not, having regard to the general area. Thus, in *Roshner v Polsue and Alfieri Ltd* (1906), a plaintiff who lived in an area which was mostly given over to printing successfully claimed that the noise of a new printing machine constituted a nuisance, as it was held that the noise of this machine was excessive even for an area largely devoted to printing. Clearly, therefore, whether a particular interference with health and comfort is actionable will depend on the exact nature of the area and the interference in question, making the chances of success at trial difficult to predict with any confidence. It should be noted, however, that the character of the neighbourhood is not relevant where property damage has been caused (*St Helen's Smelting Co v Tipping* (1865)).

Now we come to those guidelines to which the courts are ready to attach a lesser importance when the circumstances of the case demand it. First, there is the utility of the defendant's conduct, as the more useful it is, the less likely it is that the resulting interference with the claimant's land is unreasonable. This would be especially true in, for example, construction work where in addition the interference will be temporary. However, if the circumstances so require, the court will override this guideline. So, in the Irish case of *Bellew v Cement Co* (1948), the court decided that the only cement works in Ireland constituted a nuisance and granted an injunction which closed it down for a period of time, despite the fact that the supply of cement was vitally important. In *Adams v Ursell* (1913), an English court also rejected the defence that the defendant's activities were useful.

Secondly, the courts may take into account any malice on the part of the defendant. Malice is not an essential ingredient of nuisance but, if the defendant is acting maliciously, any interference caused thereby is more likely to be unreasonable. Thus, in *Christie v Davey* (1893), where the defendant's acts were totally malicious, they were held to constitute a nuisance. In *Christie*, it is quite likely that the acts of the defendants would have been held to constitute a nuisance even in the absence of malice but, in *Hollywood Silver Fox Farm v Emmett* (1936), the presence of malice converted what would probably not have been a nuisance into a nuisance. There, the defendant fired some guns at the boundary of his land adjacent to the plaintiff's land where foxes that were sensitive to noise were breeding. It was held that this constituted a nuisance, although it seems clear that in the absence of malice no nuisance would have been committed.

The final guideline to which the courts look is whether there has been some fault on the part of the defendant. Negligence is not an essential ingredient of nuisance, although it may often be present in practice, as it is no defence to an action in nuisance for the defendant to show that he took all reasonable care or even all possible care. Provided that the defendant caused (or continued) the nuisance, he is liable. However, the defendant's lack of care in allowing an annoyance to become excessive may give rise to liability in nuisance (*Andreae v Selfridge and Co* (1938)).

It can be seen from the above discussion that whether the court will decide in any particular case that the interference suffered was unreasonable is difficult to predict, and tends to support the statement which forms this question. There are, however, additional areas of uncertainty within the law of nuisance. One problem concerns who can be sued in respect of any particular interference. There is no problem where the creator of the nuisance can be identified, but problems may arise where the occupier of land from which the nuisance emanates did not create the thing which causes the nuisance. If the relevant device was created by a trespasser, the occupier will only be liable if he continues or adopts the device (*Sedleigh-Denfield v O'Callaghan* (1940)). If the occupier does neither of these things, it may be impossible to identify the trespasser, leaving the claimant without a remedy. If the nuisance arose from an act of nature, then, by the authority of *Goldman v Hargrave* (1967) and *Leakey v National Trust* (1980), the occupier must take reasonable steps to minimise foreseeable damage to others. Again, what a court will think is reasonable in any set of circumstances can be difficult to predict. If a tenant causes a nuisance on demised premises and is not worth suing because he will be unable to satisfy judgment, the landlord may be liable if he knew of the nuisance before the start of the tenancy, or if he knew the purposes for which the tenancy was created would give rise to a nuisance as an 'ordinary and necessary' consequence of the use (*Tetley v Chitty* (1986)).

It is generally not a valid defence to show that the claimant came to the nuisance (*Sturges v Bridgman* (1879)). However, in *Miller v Jackson* (1977), where some houses were built at the edge of a village green on which cricket was played, and cricket balls landed in the plaintiff's garden, Lord Denning stated that *Sturges* was no longer binding today, but this was not the view of the other members of the Court of Appeal. Lord Denning also stated in this case that where there was a conflict between public and private rights, public rights should prevail. This was exactly opposite to the view taken in the earlier case of *Pride of Derby v British Celanese* (1953), where it was held that private rights should prevail. However, in the later case of *Kennaway v Thompson* (1981), the Court of Appeal refused to follow Lord Denning's *dicta* and held that where there was a clash between private and public rights, private rights should prevail. This represents a further area of uncertainty in the law of nuisance, but it is submitted that Lord Denning's *dicta* regarding the priority of public rights do not represent the correct view of the law at present, and that his *dicta* regarding *Sturges*, a long-established case, must await confirmation by the House of Lords.

We should note that one important area of former uncertainty has been considered recently, namely, whether recovery is possible in nuisance in respect of personal injuries. In *Cunard v Antifyre* (1933), it was stated that recovery for personal injury is not possible. Also, in *Cambridge Water Co v Eastern Counties Leather plc* (1994), the House of Lords, when considering an action under the rule in *Rylands v Fletcher* (1868), referred with approval to a 'seminal' article by Professor Newark ('The boundaries of nuisance' (1949) 65 LQR 480), in which he argued that recovery for personal injury should not be possible in nuisance. Although the House did not decide the situation regarding recovery in respect of personal injury in *Rylands*, let alone in nuisance, their Lordships' wholehearted acceptance of Professor Newark's article suggests that a future court will not allow such recovery. Thus, it came as no surprise when the House of Lords in *Hunter v Canary Wharf Ltd* (1997) followed their reasoning in *Cambridge Water Co* and declared that actions for personal injury should not be brought in nuisance.

Another debatable point is whether economic loss can be recovered in nuisance. Although there exist *dicta* in *British Celanese v Hunt* (1969) and *Ryeford Homes v Sevenoaks District Council* (1989) which suggest that economic loss is recoverable, the whole tenor of the judgments of the House of Lords in *Cambridge Water Co* and *Hunter* is against such recovery.

It should also be noted that if the activity which it is claimed constitutes a nuisance is regulated by statute, a common law claim in nuisance will not be allowed if it is inconsistent with the statutory scheme (*Marcic v Thames Water Utilities Ltd* (2004)).

Finally, we should note the recent and possibly far-reaching effect of the Human Rights Act 1998 on the law of nuisance. Article 8 of the European Convention on Human Rights, brought into the law by s 1 of the 1998 Act, establishes the right to respect for private and family life and home. Article 1 of the First Protocol states persons are entitled to the peaceful enjoyment of their possessions and Art 2 of the Convention establishes a right to life.

As regards Art 8, in *Lopez Ostra v Spain* (1995), it was held that the construction of a waste treatment plant next to the applicant's house, which had caused local pollution and health problems, was a violation of Art 8. In this case the Spanish Government did not own the plant, but it was held to be sufficient that the local authority had allowed it to be built on their land and the Spanish Government had subsidised it. As s 6(1) of the 1998 Act makes it unlawful for a public authority to act in a way incompatible with a Convention right, and by s 6(6) an 'act' includes a failure to act, both the Government and local authorities could be held liable for breaches of Art 8. Article 8 has also been held to cover noise (*Baggs v UK* (1987); *Hatton v UK* (2003)) and toxic emissions (*Guerra v Italy* (1998)). Clearly, hazardous emissions could fall within Art 1 and even Art 2 if the emissions are sufficiently hazardous.

This new jurisprudence could have extensive effects on the law of nuisance. An action under the Human Rights Act 1998 would raise no problems of interest in land, recovery of economic loss or application to personal injuries. Indeed, in *Marcic* (2001), in the High Court, the judge found for the plaintiff under Art 8 while dismissing the claims based on nuisance and *Rylands*. Although the House of Lords dismissed the plaintiff's claims in both nuisance and under Art 8, it is clear from the judgments in both the High Court and the Court of Appeal that much of the detailed law of nuisance is irrelevant in considering a breach of Art 8.

However, the European Convention on Human Rights does not automatically provide a mechanism for bypassing much of the detailed common law of nuisance. In *Hatton* and *Marcic*, the fundamentally subsidiary nature of the Convention was emphasised. In *Hatton*, it was stated that national authorities have direct democratic legitimation and are well-placed to evaluate local needs and conditions. A fair balance must be struck between the interests of the individual and the interests of the community as a whole. Thus, if the appropriate national or local authority consults widely and considers carefully all responses to its consultation, it will not be found in breach of the Convention (*Hatton*).

A further restriction on the use of the Convention is the existence of a detailed statutory regime in which the potential nuisance exists. Thus, in *Marcic*, the House of

Lords held the contents of a statutory scheme for a statutory sewerage undertaker struck a reasonable balance between the needs of the individual and of the community. In *Hatton*, it was held that in matters of general policy, on which opinions might differ widely, the role the domestic policymaker should be given special weight. So, given the fundamentally subsidiary nature of the Convention, Art 8 could not override the statutory scheme.

Thus, taking an overall view of the law of nuisance, we can see that, despite several recent decisions which have introduced greater certainty into the law of nuisance, there are a number of areas where either the law is uncertain, or where it would be difficult to predict with any confidence at all what decision a court would come to, faced with a particular set of facts, whether the common law of nuisance or the European Convention on Human Rights is invoked.

Notes

9 The Rule in *Rylands v Fletcher* and Fire

Introduction

Questions on *Rylands v Fletcher* (1868) are popular with examiners as there are a number of undecided aspects to the rule, and because it is very easy to combine a *Rylands* situation with elements of nuisance, negligence or animals.

Checklist

Students must be familiar with the following areas:

(a) the elements of the rule itself, with especial reference to:

- the non-natural user requirement;
- the recent decision of the House of Lords in *Transco v Stockport MBC* (2004); and

(b) defences, and especially the independent acts of third parties.

Question 29

Delta Manufacturing plc owns and operates a factory situated on an industrial estate on the outskirts of a small town. One day, the environmental control system malfunctioned for some unknown reason and large quantities of toxic fumes were emitted. These fumes damaged paintwork on some houses in the town and some inhabitants also suffered an allergic reaction to the fumes. As a result of the adverse publicity, the town has seen a reduction in its normal tourist trade and the local shopkeepers are complaining of loss of business.

Advise Delta Manufacturing plc of any liability it might have incurred.

Would your advice differ if Delta operated its factory under statutory authority?

Answer plan

It is important in answering this question to consider the possible courses of action in detail, paying particular attention to *Rylands* and nuisance, and the possibility of a negligence action. The defence of statutory authority must also be considered for these actions.

The following points need to be discussed:

- ingredients of *Rylands v Fletcher* (1868) with especial reference to non-natural use;
- recoverability for property damage by landowners and non-landowners;
- ingredients of nuisance;
- negligence and the problem of proof of breach of duty;
- statutory authority as a defence to the above actions; and
- action under the Human Rights Act 1998.

Answer

We shall first consider whether Delta has incurred any liability under the rule in *Rylands v Fletcher* (1868), which is that a 'person who, for his own purposes, brings onto his lands and collects and keeps there anything likely to do mischief if it escapes must keep it in at his peril, and if he does not do so, he is *prima facie* answerable for all the damage which is the natural consequence of its escape'. In addition, the defendant must have made a 'non-natural' use of his land, and the harm caused must be foreseeable (*Cambridge Water Co v Eastern Counties Leather plc* (1994)). The whole area of *Rylands* has been reconsidered recently by the House of Lords in *Transco v Stockport Metropolitan Borough Council* (2004) and we shall consider the effect of this case on Delta's liability.

The fumes have been brought onto Delta's land for Delta's purposes. They have been brought onto Delta's land in the sense that they are not something that is there by nature, such as thistles (*Giles v Walker* (1890)) or rainwater (*Smith v Kenrick* (1849)). The toxic fumes are clearly likely to do mischief if they escape, and there has been an escape from Delta's premises as required by *Read v Lyons* (1947). As it is foreseeable that the fumes would cause harm, we must determine whether or not there has been a non-natural use of land, an aspect that has given rise to much confusion. In *Rylands* itself, the word 'natural' was used to mean something on the land by nature, but later cases have construed the word as meaning 'ordinary' or usual. In *Rickards v Lothian* (1913), Lord Moulton said of the use of land required to bring *Rylands* into operation: 'It must be some special use bringing with it increased danger to others, and must not merely be the ordinary use of the land or such a use as is proper for the general benefit of the community.' In *Read*, Viscount Simon described Lord Moulton's analysis of *Rylands* as 'of the first importance'. In *Transco*, the House of Lords considered in detail the non-natural requirement. Lord Bingham stated that the ordinary use test is to be preferred to the non-natural use test as this makes it clear that *Rylands* only applies where the use of land is extraordinary and unusual. Lord Bingham doubted that a test of reasonable use was helpful since a use may be out of the ordinary but reasonable, for example, the storage of chemicals on industrial premises as in *Cambridge Water*. It was also doubted that Lord Moulton's criterion of whether the use is proper for the general benefit of the community was helpful, echoing the criticism of this phrase by Lord Goff in *Cambridge Water*. Thus Lord Bingham stated that it was necessary to show that the defendant had brought onto or kept on his land an exceptionally dangerous or mischievous thing in extraordinary or unusual circumstances, and Delta have done.

If the rule were to be applicable, then the houseowners could recover for damage to their paintwork (*Rylands*). However, houseowners could not recover for their allergic reaction, as the rule in *Rylands* does not cover personal injury: *Transco*. If any person suffered personal injury but had no interest in land, that lack of interest would itself rule out an action in *Rylands*: *McKenna v British Aluminium Ltd* (2002). The local shopkeepers have suffered economic loss and, despite *Weller v Foot and Mouth Disease Research Institute* (1965), there seems to be no clear authority for recovery on their part, and the general tenor of *Cambridge Water Co* (and the analogous case of *Hunter*) is against such recovery.

We shall next consider whether any action will lie against Delta in nuisance. There would be no problem to houseowners or tenants recovering for the damage to their paintwork. As this involves damage to property, the character of the neighbourhood is not a relevant factor (*St Helen's Smelting Co v Tipping* (1865)), although the persons affected will have to show that they have an interest in the land affected (*Malone v Laskey* (1907); *Hunter v Canary Wharf Ltd* (1997)). Even persons with an interest in land, however, will be unable to claim for any allergic reactions as, in *Hunter*, the House of Lords stated that actions for personal injury should not be brought in nuisance. Although there are *dicta* in *British Celanese v Hunt* (1969) and *Ryeford Homes v Sevenoaks District Council* (1989) which suggest that recovery for economic loss is possible, the tenor of the judgment of the House of Lords in *Hunter* is against such recovery.

An action in public nuisance may also lie against Delta. Here, the claimant will have to show that the nuisance affected a section of the public (*AG v PYA Quarries* (1957)) and that he suffered damage over and above that suffered by the public at large. The advantage to claimants in public nuisance is that no interest in land is required and both personal injury and economic loss are recoverable (*Rose v Miles* (1815)). Thus, those persons with no interest in land could sue in respect of the allergic reaction, which would constitute special damage, as could the shopkeepers.

Also, Delta may be liable in negligence. There would be no difficulty in showing the existence of a duty of care and causation and foreseeability, but there could be problems in proving breach, as we are told that the emission occurred for an unknown reason. A possible claimant might seek to rely on *res ipsa loquitur*, but this would not reverse the burden of proof, which lies on the claimant throughout (*Ng Chun Pui v Lee Chuen Tat* (1988)). If Delta could show that it had in place a proper system of inspection and control (*Henderson v Jenkins and Sons* (1970)), this would be sufficient to negate liability. If negligence could be proved against Delta, then of course any claimant who has suffered damage to property or to the person may sue, but the shopkeepers would be unable to recover for their economic loss, as the chances of a claimant now successfully relying on *Junior Books v Veitchi* (1983) seem non-existent.

If the factory had been operated under statutory authority, liability would not arise either under *Rylands* or nuisance unless negligence on the part of Delta could be shown (*Green v Chelsea Waterworks* (1894); *Allen v Gulf Oil Refining* (1981)).

Finally, we must consider any causes of action that might arise under the Human Rights Act 1998. Article 8 of the European Convention on Human Rights, brought into UK law by s 1 of the 1998 Act, establishes the right to respect for private and family life. Also, Art 1 of the First Protocol states that persons are entitled to the peaceful enjoyment of their possessions, and Art 2 establishes a right to life.

As regards Art 8, in *López Ostra v Spain* (1995), it was held that the construction of a waste treatment plant next to the applicant's house, which had caused local pollution and health problems, was a violation of Art 8. In this case, the Spanish Government did not own the plant, but it was held to be sufficient that the local authority had allowed it to be built on their land and the Government had subsidised it. As s 6(1) of the 1998 Act makes it unlawful for a public authority to act in a way incompatible with a Convention right, and by s 6(6) an 'act' includes a failure to act, both the UK Government and local authorities could be held liable for breaches of Art 8. Article 8 has been held to cover toxic emissions from a factory (*Guerra v Italy* (1998)), so clearly the emissions from Delta's factory would fall within Art 8 and even Art 2 if the emissions are sufficiently hazardous.

An action under the Human Rights Act 1998 would raise no problems as regards interest in land, recovery for economic loss or recovery for personal injuries. Indeed, in *Marcic v Thames Water Utilities Ltd* (2001), the High Court found for the claimant under Art 8 while dismissing the claims based on nuisance and *Rylands*. Although the Court of Appeal (2002) held that the claimant could recover in nuisance while upholding the High Court's findings under the Human Rights Act 1998, it is clear from the High Court decision that much of the detailed law on nuisance and *Rylands* is irrelevant in considering a breach of Art 8. Thus, an action under the Human Rights Act 1998 would be available to a considerable range of persons who suffer injury due to Delta's factory, and the possible defendants could be the local authority under s 6(1) and (6) of the 1998 Act, or the UK Government.

Question 30

'Although the rule in *Rylands v Fletcher* has been subject to recent judicial scrutiny, there still remain areas of uncertainty and it is doubtful if it adds anything to existing English law.'

Discuss the above statement.

Answer plan

This is a general essay question requiring a discussion of the similarity and differences between *Rylands v Fletcher* (1868) and nuisance, especially since the decisions of the House of Lords in *Cambridge Water Co v Eastern Counties Leather plc* (1994), *Hunter v Canary Wharf* (1997) and *Transco v Stockport Metropolitan Borough Council* (2004).

The following aspects need to be discussed:

- ingredients of an action in *Rylands*;
- similarity with an action in nuisance;
- problems raised by the requirement in *Rylands* for an escape and non-natural use; and
- other actions which may reinforce *Rylands*, for example, animals, trespass or negligence.

Answer

In *Rylands v Fletcher* (1868), Blackburn J gave the classic statement of the law when he stated: 'We think that the true rule of law is that the person who, for his own purposes, brings onto his land and collects and keeps there anything likely to do mischief if it escapes, must keep it in at his peril, and, if he does not do so, he is *prima facie* answerable for all the damage which is the natural consequence of its escape.' This statement was approved when the case was appealed to the House of Lords, where Lord Cairns LC made the crucial addition that the defendant also had to make a 'non-natural' use of his land.

The scope and certainty of the rule can be considered under the following headings.

Accumulation

The rule refers to 'bringing' things onto the defendant's land, and thus does not apply to things which are naturally on the land, such as thistles (*Giles v Walker* (1890)) or rainwater (*Smith v Kenrick* (1849)).

Dangerous things

The rule refers to 'anything likely to do mischief if it escapes'. However, there are very few objects which do not give rise to some risk if they escape. In *Transco v Stockport Metropolitan Borough Council* (2004), the House of Lords stated that the mischief test should not be easily satisfied. It should be shown that the defendant had done something he recognised or ought to have recognised as giving rise to an exceptionally high risk of danger or mischief if it escapes, however unlikely such an escape may have been thought to be. In *Read v Lyons* (1947), Lord Macmillan stated that it would not be practicable to classify objects into dangerous and non-dangerous things. It is no longer considered necessary that the thing be dangerous, but danger is still relevant when considering non-natural use or foreseeability of damage.

Escape

The rule clearly states that the thing must escape from the defendant's land, and the necessity for this was emphasised by the House of Lords in *Read v Lyons* (1947) and in *Transco*.

His land

Although the rule refers to 'his land', there is no requirement that the defendant be the owner of the land – it would seem from the cases that it is enough that the defendant has control of the thing. This is similar to the position in nuisance and, indeed, the House of Lords in *Cambridge Water Co v Eastern Counties Leather plc* (1994) held that the rule in *Rylands* is basically the law of nuisance extended to cover an isolated escape.

His own purposes

This requirement suggests that if the defendant brings the thing onto his land for some other person's purpose, the rule ceases to apply. This is often said to be supported by the decision of the House of Lords in *Rainham Chemical Works v Belvedere Fish Guano Co* (1921), although in that case liability was admitted at first instance and the appeals to the Court of Appeal and the House of Lords were concerned solely with whether the directors of the company could be held personally liable. Thus, *Rainham* is very questionable support for this requirement.

Non-natural use

The original meaning of the phrase 'natural use' was something that was there naturally or by nature, but gradually the courts interpreted it to mean 'ordinary' or 'usual'. In *Rickards v Lothian* (1913), Lord Moulton stated: 'It must be some special use bringing with it increased danger to others, and must not merely be the ordinary use of land or such use as is proper for the general benefit of the community.' Although this was described in *Read* by Viscount Simon as 'of the first importance', it was criticised by the House of Lords in *Cambridge Water Co*. Lord Goff stated that the phrase 'ordinary use of land' was lacking in precision, and that the alternative criterion 'or such as is proper for the general benefit of the community' introduced doubt and might not keep the exception within reasonable bounds. In *Transco*, the House of Lords considered in detail the non-natural requirement. Lord Bingham stated that the ordinary use test is to be preferred to the natural use test, as this makes it clear that *Rylands* only applies where the use of land is extraordinary and unusual. Lord Bingham doubted that a test of reasonable use was helpful since a use may be out of the ordinary but reasonable, for example, the storage of chemicals on industrial premises as in *Cambridge Water*. It was also doubted that Lord Moulton's criterion of whether the use is proper for the general benefit of the community was helpful, echoing the criticism of this phrase by Lord Goff in *Cambridge*. Thus, Lord Bingham stated that it is necessary to show that the defendant had brought an exceptionally dangerous or mischievous thing in extraordinary or unusual circumstances.

Foreseeability of damage

In *Cambridge Water Co*, after a thorough historical survey of the rule, it was held that foreseeability of damage following escape was a necessary ingredient of an action under the rule in *Rylands*.

Damage covered

Property damage is clearly covered by the rule, but personal injuries are excluded as personal injuries do not relate to any right in or enjoyment of land: *Transco*.

Interest in land

Because of the similarity of *Rylands* and nuisance (*Cambridge Water Co*), it has been held that claimants under *Rylands* must have a proprietary interest in the land affected: *McKenna v British Aluminium* (2002).

It can be seen that *Rylands* bears a close resemblance to nuisance. Nuisance may be defined as an unreasonable interference with a person's use or enjoyment of land, or some right over it or in connection with it. It does not require an accumulation, as it applies, for example, to noise, and it applies to both dangerous and non-dangerous things. The relevance of the thing being dangerous is as to whether the defendant has made a reasonable use of his land. Nuisance differs from *Rylands*: in *Rylands*, there are defined ingredients to the tort; in nuisance, there are guidelines as to whether the interference with the claimant's land was unreasonable. Thus, in nuisance, the court will take into account the duration of the interference, whether it was of a temporary nature and whether it was an isolated event. It was held in *Bolton v Stone* (1951) that an isolated happening could not constitute a nuisance, whereas in *Cambridge Water Co*, it was held that such an isolated event could found an action under *Rylands*.

By the very nature of nuisance, the thing, be it noise or a physical thing, must escape from the defendant's land. Also in nuisance, there is no requirement that the defendant be the owner of the land, mere control being sufficient. There is no requirement in nuisance that there is a non-natural use of the land, only that it is unreasonable. It is, of course, possible that a natural use of land will be unreasonable due to (say) the presence of malice on the part of the defendant (*Hollywood Silver Fox Farm v Emmett* (1936)). In both torts, foreseeability of damage is required, and neither of these torts cover personal injuries.

Despite the relaxation of the requirement that the claimant has an interest in land by the Court of Appeal in *Khorasandjian v Bush* (1993) and *Hunter v Canary Wharf* (1996), the House of Lords reinstated this requirement in *Hunter* (1997).

Thus, it can be seen that there is an overlap between *Rylands* and nuisance and in many situations the two causes of action may co-exist. *Rylands* does, however, fill one gap in the law, in that it does apply to an isolated event whereas nuisance does not. It has also been held that nuisance does not cover interference with purely recreational matters (*Bridlington Relay v Yorkshire Electricity Board* (1965)), whereas this restriction does not apply to *Rylands*.

It could thus be said that in practice, the majority of cases in which *Rylands* applies will also give rise to causes of action in nuisance and possibly other torts such as negligence, animals or trespass. However, *Rylands* does cover some areas that other torts do not cover, such as the isolated event which is not covered in nuisance, and the isolated event caused by the action of an independent contractor, which would be covered in neither nuisance nor negligence. It therefore, to this extent, adds to existing English law even though, as noted in *Transco*, no claimant has succeeded in a *Rylands* action in the past 60 years.

It can also be seen that recent cases have brought an element of certainty into the law regarding *Rylands*. It is now clear that foreseeability of damage is an essential ingredient, as is an interest in land, and that *Rylands* is not applicable to personal injuries. However, the discussion of the phrase 'natural use' in *Transco* and the use of the phrase 'extraordinary and unusual' use does not seem to have introduced much certainty into this area. Similarly, the suggestion that the 'mischief' criterion should not be at all easily satisfied and that the test that the defendant has done something which he recognised or ought to have recognised as giving rise to an exceptionally high risk of danger or mischief if it escapes seems also to lack certainty.

Notes

Question 31

Edward owns a garden centre in a rural area. He specialises in growing and selling orchids which need to be reared in heated glasshouses. He has an extremely large storage tank containing heating oil which he uses to heat the glasshouses. Due to internal corrosion of the tank, the oil escapes and contaminates some vegetables growing on a farm belonging to Frank, Edward's neighbour. The oil also escapes onto the road and Frank, who is driving along the road at the time, skids and crashes his car, and as a result suffers a cut to his head.

Advise Edward.

Would your advice differ if the escape of oil had been caused by Jack, a rival of Edward's, opening the tap of the oil tank?

Answer plan

This is another question involving a multiplicity of causes of action, namely, *Rylands v Fletcher* (1868), nuisance and negligence, and the ingredients and defences to these actions must be considered.

The following points in particular need to be discussed:

- ingredients of *Rylands,* and especially non-natural user;
- types of damage recoverable under *Rylands*;
- ingredients of nuisance;
- damages recoverable under nuisance;
- liability in negligence; and
- act of third party as defence to *Rylands*: nuisance, negligence.

Answer

We shall first consider whether Edward has incurred any liability under the rule in *Rylands v Fletcher* (1868), which is that a person who, for his own purposes, brings onto his land and collects and keeps there anything likely to do mischief if it escapes must keep it in at his peril, and if he does not do so, he is *prima facie* answerable for all the damage which is the natural consequence of its escape. In addition, the defendant must have made a non-natural use of his land.

The oil has been brought onto Edward's land for Edward's purposes, and has been brought onto his land in the sense that it is not something that is there by nature, such as thistles (*Giles v Walker* (1890)) or rainwater (*Smith v Kenrick* (1849)). The oil is clearly likely to do mischief if it escapes, and there has been an escape from Edward's land as required by *Read v Lyons* (1947). Finally, we must determine whether or not there has been a non-natural use of land, an aspect that has given rise to much confusion. In *Rylands* itself, the word 'natural' was used to mean something on the land by nature, but later cases have construed the word as meaning 'ordinary' or 'usual'. In *Rickards v Lothian* (1913), it was said that a non-natural use was some special use bringing with it increased danger to others, and not merely a use which brings general benefits to the community. In *Transco v Stockport Metropolitan Borough Council* (2004), the House of Lords considered in detail the non-natural requirement. Lord Bingham stated that the ordinary use test is to be preferred to the natural use test as this makes it clear that *Rylands* only applies where the use of land is extraordinary and unusual. He doubted that a test of reasonable use was helpful since a use may be out of the ordinary but reasonable, for example, the storage of chemicals on industrial premises as in *Cambridge Water Co v Eastern Counties Leather plc* (1994). It was also doubted that Lord Moulton's criterion in *Rickards v Lothian* (1913) of whether the use is proper for the general benefit of the community was helpful, echoing the criticism of this phrase by Lord Goff in *Cambridge Water*. Thus, Lord Bingham stated that it is necessary to show that the defendant had brought or kept on his land an exceptionally dangerous or mischievous thing in extraordinary or unusual circumstances, and this Edward has done.

Turning now to the damage caused, it is clear from *Rylands* itself that Frank can recover for the damage to his vegetables, provided that he has an interest in the land upon which the vegetables are being grown. In *McKenna v British Aluminium Ltd* (2002), it was held that as the House of Lords in *Cambridge Water Co* had stated that *Rylands* was effectively an extension of the law of nuisance, any claimants under *Rylands* would have to show a proprietary interest in the land affected. As regards the damage to Frank's car, Frank is a landowner who has suffered property damage, but at the time of the damage the property was not on his land. In *Halsey v Esso Petroleum* (1961), the plaintiff was allowed to recover under these circumstances. The cut to Frank's head is damage to the person and, in *Transco*, the House of Lords held that *Rylands* does not apply to personal injuries.

We must next consider whether Edward has incurred any liability in nuisance. A nuisance is an unreasonable interference with a person's use or enjoyment of land, or some right over or in connection with it. However, not all interference necessarily gives rise to liability, and there must be give and take between neighbours. Also, the interference must be substantial and not fanciful (*Walter v Selfe* (1851)). The courts have developed a number of guidelines that are used to determine whether any particular interference is unreasonable, but each test is only a guideline and not a condition, and the court has to evaluate the defendant's behaviour in all the circumstances of the case. In Edward's case, the court will consider whether the escape was an isolated event. In *Bolton v Stone* (1951), it was stated that a nuisance could not arise from an isolated happening, but that it had to arise from a state of affairs, however temporary. Thus, in *Midwood v Manchester Corp* (1905), a gas explosion was held to be a nuisance, because although it was an isolated event, it was due to a pre-existing state of affairs, namely, the build-up of gas. On this basis, it could be argued that the escape of oil was due to a build-up of this material on Edward's premises, and thus the escape can constitute an actionable nuisance. The damage suffered is not due to any sensitive use of the property by the neighbour (as was the case in *Robinson v Kilvert* (1889)), and the character of the neighbourhood is not to be taken into account where physical damage to property has been caused (*St Helen's Smelting Co v Tipping* (1865)). Note that it would not be necessary to show that Edward was negligent, as negligence is not an essential ingredient of nuisance. Indeed, it would be no defence for Edward to show that he took all reasonable care or even all possible care – provided that he caused the nuisance, that is sufficient. Thus, taking all the circumstances into account, a court would find that the escape of oil constituted an actionable nuisance.

Again, considering the damage caused, Frank can recover for the damage caused to his vegetables. Following *Cambridge Water Co,* however, he cannot recover for his personal injuries. As regards the damage to his car, Frank has the problem that he has no interest in the road, which is a prerequisite to recovery in nuisance (*Cambridge Water Co*). However, Frank could sue in public nuisance, as the presence of the oil on the road would affect a section of the public (*AG v PYA Quarries* (1957)), and Frank has suffered damage over and above that suffered by the public at large (*Rose v Miles* (1815)). In public nuisance, the claimant need have no interest in land and can recover for personal injury, so Frank could recover for the damage to his car and for the cut to his head.

Frank could also sue Edward in negligence, and would have no difficulty in establishing a duty of care, causation and foreseeability of damage. A problem might arise, however, with breach of duty, as we are told that the leak arose from internal corrosion. Frank might seek to rely on *res ipsa loquitur,* but this would not reverse the burden of proof which lies on the claimant throughout (*Ng Chun Pui v Lee Chuen Tat* (1988)). If Edward could show that he had in place a proper system of inspection and control, that would be sufficient to negate liability (*Henderson v Jenkins and Sons* (1970)).

If negligence could be proved against Edward, Frank could recover for all the damage that he has suffered, as there seem to be no problems in causation and remoteness.

If the leak had been caused by the deliberate action of Jack, that would provide a defence to Edward in an action under *Rylands*. In *Rickards v Lothian* (1913), it was held that the defendants were not liable because the cause of the damage was an unforeseeable independent act of a third party over whom the defendant has no control (see also *Perry v Kendricks Transport* (1956)). In nuisance, the occupier is liable only if the damage is foreseeable (*Hunter v Canary Wharf* (1997)), which it does not appear to be in these circumstances. In addition, where a nuisance is caused by the act of a trespasser, the occupier is only liable where he continues or adopts the nuisance (*Sedleigh-Denfield v O'Callaghan* (1940)). As Edward has neither adopted nor continued the nuisance, he would not be liable in nuisance for the action of Jack. Similarly, in negligence, Edward would be under no duty of care to prevent Jack's action (*Smith v Littlewoods Organisation* (1987)) and would not be liable for any damage flowing from such an action.

Notes

Question 32

One evening, Henry lights a bonfire in his garden in order to burn some garden rubbish. The smoke and smell from the bonfire annoy his neighbours who are watching television with the windows open, and sparks from the fire damage some clothing that one of his neighbours has hung out in his garden to dry. The smoke from the bonfire drifts onto the road and is so thick that it obstructs the vision of a passing motorist who as a result runs into a lamp post. Henry goes indoors to listen to the radio, and some time later the bonfire spreads to his neighbour's property and destroys a garden shed.

Advise Henry of his legal liability.

Answer plan

This is a question that requires a discussion of Henry's liability in nuisance, the relationship of nuisance to an action in *Rylands v Fletcher* (1868), and any liability Henry might incur in negligence and under the special rules that govern fires.

The following points need to be discussed:

- liability in nuisance for the smoke and smell;
- liability in nuisance for the damage to the clothing;
- possibility of liability arising under the rule in *Rylands v Fletcher* (1868);
- liability in negligence; and
- liability for the fire under the Fires Prevention (Metropolis) Act 1774.

Answer

We shall first consider any liability that Henry may have incurred in private nuisance (which we shall henceforth simply refer to as nuisance) for the smoke and smell from his bonfire. A nuisance consists of an unreasonable interference with a person's use or enjoyment of land, or of some right over or in connection with it. However, not all interference will necessarily give rise to liability: the harm must be foreseeable (*The Wagon Mound (No 2)* (1967); *Cambridge Water Co v Eastern Counties Leather plc* (1994)), and the interference must be substantial and not merely fanciful (*Walter v Selfe* (1851)). In deciding whether a particular interference is unreasonable or not, the court will rely on a series of guidelines rather than on any rigid rules. In Henry's case, the court would consider the duration of the interference, as the shorter the duration of the interference, the less likely it is to be unreasonable, as in *Harrison v Southwark and Vauxhall Water Co* (1891). In particular, it seems that an isolated event is unlikely to constitute a nuisance. In *Bolton v Stone* (1951), it was stated that a nuisance must be a state of affairs, however temporary, and not merely an isolated happening. Thus, although Henry might claim that the bonfire is an isolated event, it does constitute a temporary state of affairs, and is capable in law of being a nuisance. A possible argument that Henry might employ is that he only lights a bonfire on rare occasions and that this is a reasonable use of his land. However, the fact that a defendant is only making reasonable use of his land is not, of itself, a valid defence in nuisance (*AG v Cole* (1901); *Vanderpant v Mayfair Hotel* (1930)). As regards any interference with health and comfort, the court will take into account the character of the neighbourhood, as 'what would be a nuisance in Belgravia Square would not necessarily be so in Bermondsey' (*Sturges v Bridgman* (1879), *per* Thesiger LJ). Thus, if Henry lives in a suburban or rural area, the occasional lighting of a bonfire might not constitute a nuisance, as there must be an element of give and take between neighbours. However, if Henry by his lack of care allowed an annoyance from the bonfire to become excessive, he would become liable in nuisance (*Andreae v Selfridge and Co* (1938)). Hence, as regards the smoke and smell from his bonfire, whether Henry will be liable in nuisance will depend on whether, taking all the circumstances into account, the interference is unreasonable. As nuisance protects a person's use or enjoyment of land, then traditionally only those neighbours with an interest in the land can sue (*Malone v Laskey* (1907)). Despite the relaxation of this requirement by the Court of Appeal in *Khorasandjian v Bush* (1993) and *Hunter v Canary Wharf Ltd* (1996), the House of Lords reinstated the requirement when it heard *Hunter* (1997). Thus, only those neighbours with an interest in the property affected (for example, houseowners or tenants) can sue, and not merely members of their families or guests. It was also held in *Bridlington Relay v Yorkshire Electricity Board* (1965), and confirmed by the House of Lords in *Hunter* (1997), that interference with purely recreational facilities, such as television reception, would not constitute an actionable nuisance. However, the interference suffered by Henry's neighbours is not with the reception of their television programmes, but rather with their enjoyment of their property, for had they wished to just sit in their houses with the windows open, they would not have been able to do so without the discomfort from the smoke and smell of Henry's bonfire.

Turning now to the damage to the neighbour's clothing, where physical damage to property has been caused, the character of the neighbourhood is not relevant (*St Helen's Smelting Co v Tipping* (1865)), and a court would be far more likely to find that an interference is unreasonable where physical damage to property has occurred. Even if the bonfire did not originally constitute a nuisance, Henry's lack of care in allowing the interference to become unreasonable would make him liable (*Andreae*). It therefore seems likely that Henry would be liable for the damage to his neighbour's clothing, providing of course that his neighbour has the required interest in land. Henry could also incur liability for the damage to his neighbour's clothing in negligence. Henry will owe his neighbour a duty of care under normal *Donoghue v Stevenson* (1932) principles. As a duty of care has already been held to exist in such circumstances, there is no need to go to the modern incremental formulation of the test for a duty of care that was preferred by the House of Lords in *Caparo Industries plc v Dickman* (1990) and *Murphy v Brentwood District Council* (1990). In allowing sparks to damage his neighbour's property, Henry has not acted as a reasonable person would, and so is in breach of his duty (*Blyth v Birmingham Waterworks* (1856)), and the 'but for' test of Lord Denning in *Cork v Kirby MacLean* (1952) shows the required causal connection. Finally, the damage suffered by the neighbour is not too remote as it is reasonably foreseeable (*The Wagon Mound (No 1)* (1961)). Thus, Henry would be liable for the damage to the clothing, and there would be no requirement in negligence for the neighbour to have any interest in land.

As regards the passing motorist, he could not sue Henry in nuisance, as he has no interest in the land. He could sue Henry in negligence, as the required elements of duty, breach and damage appear to be present (see the above discussion regarding the neighbour and his damaged clothing). The motorist may also have a cause of action in public nuisance, in that Henry has created a danger close to the highway (*Tarry v Ashton* (1876); *Castle v St Augustine's Links* (1922)).

We shall next consider whether Henry has incurred any liability for the fire and the damage it has caused to the garden shed. Liability could arise in a number of ways: the first possibility is an action under the rule in *Rylands v Fletcher* (1868). However, in *Mason v Levy Auto Parts* (1967), MacKenna J held that liability for fire cannot be based on *Rylands* because the 'thing' has not escaped from the defendant's land as required by *Rylands*; see also *Johnson v BJW Property Developments Ltd* (2002). Instead, Henry may be liable under common law liability for fire. Here the claimant will have to show: first, that Henry brought onto his lands things likely to catch fire, and kept them there in such condition that if they did ignite the fire would be likely to spread to the claimant's land; secondly, that he did so in the course of some non-natural use of the land; and finally, that the things ignited and the fire spread. Although these are different criteria from those used in *Rylands*, similar considerations will apply in deciding whether these criteria have been satisfied in any particular case. The only element that would appear to give rise to any problems here is the requirement that the use of land be non-natural. In *Rylands* itself, the word 'natural' was used to mean something that was there by nature. However, in *Rickards v Lothian* (1913), Lord Moulton stated that a non-natural use must be 'some special use bringing with it increased dangers to others, and must not merely be the ordinary use of the land or such a use as is proper for the general benefit of the community'. In *Transco v Stockport Metropolitan Borough Council* (2004), the House of Lords considered in detail the non-natural use requirement. Lord Bingham stated that the ordinary use test is to be preferred to the non-natural use test as this makes it clear that *Rylands* only applies where the use of land is extraordinary and unusual. He doubted that a test of reasonable use was helpful since a use may be out of the ordinary but reasonable, for example, the storage of chemicals on industrial premises as in *Cambridge Water v Eastern Counties Leather plc* (1994). It was also doubted that Lord Moulton's criterion of whether the use is proper for the general benefit of the community was helpful, echoing the criticism of this phrase by Lord Goff in *Cambridge Water*. Thus, Lord Bingham stated that it is necessary to show that the defendant had brought or kept on his land an exceptionally dangerous or mischievous thing in extraordinary or unusual circumstances, and this Henry has done. Henry would thus be liable in a common law action for fire. In addition, Henry would also incur liability at common law in nuisance, as the fire has damaged his neighbour's property (*Goldman v Hargrave* (1967)), assuming that his neighbour has the necessary interest in the land. Liability could also attach in negligence, as there is no problem in establishing a duty of care, causation and damage that is not too remote and, by leaving the fire to go indoors and listen to the radio, Henry has failed to take reasonable care to prevent the fire from causing damage (*Musgrove v Pandelis* (1919); *Ogwo v Taylor* (1987)). Now we must consider whether Henry could escape liability by relying on the provisions of the Fires Prevention (Metropolis) Act 1774. Section 86 of that Act provides (in archaic language) that no action shall be brought or damages recovered in respect of a fire which starts accidentally. So Henry will not be liable for the consequences of the fire if it began accidentally. The meaning of 'accidentally' was considered in *Filliter v Phippard* (1847), where the defendant deliberately lit a fire to burn some weeds and then neglected the fire, which spread to the plaintiff's land and damaged his hedge. It was held that the defendant could not rely on the Act, because the fire did not begin 'accidentally' – it began negligently. The court held that a fire only began accidentally where it began by mere chance or was incapable of being traced to any cause (see also *Johnson*). As *Filliter* is legally indistinguishable from Henry's situation, it follows that Henry cannot rely on the 1774 Act as a defence. In *Johnson* it was held that 'accidentally' applied to the escape of fire rather than the manner in which the fire started, but even on this interpretation Henry would not be able to rely on the 1774 Act.

Thus, Henry should be advised that he will be liable for the damage to the clothing and to the shed, and for the damage suffered by the motorist.

Notes

10 Animals

Introduction

Questions involving animals may arise in examinations in a number of ways. A question whose main ingredient is nuisance or *Rylands v Fletcher* (1868) or negligence may involve animals, but we are concerned in this chapter with questions where the topic being tested is mainly the Animals Act 1971 and related common law issues.

Checklist

To attempt a question on animals, students must be familiar with the following areas:

(a) the common law situation;

(b) definition of a dangerous species;

(c) liability for damage caused by dangerous and non-dangerous species and the decision of the House of Lords in *Mirvahedy v Henley*;

(c) defences;

(d) definition of a keeper of an animal; and

(e) straying livestock.

Question 33

Graham owns a large Alsatian dog, which he lets roam in his garden to deter unwelcome visitors. One day, the dog jumps over the low garden fence to chase a cat and the cat runs into the road to escape and is run over. Helen, who owns the cat, is told of this incident by a neighbour who witnessed it and, later that evening, Helen goes to Graham's house to demand compensation for her cat. Before she can enter Graham's garden, the dog jumps over the fence and bites Helen, who in an attempt to escape further attack runs into the road. Fiona, who is driving along the road at the time, swerves to avoid Helen, runs into a lamp post and is injured.

Advise Graham of any liability that may have arisen.

Answer plan

This is a relatively straightforward question (though the position with Helen's cat is rather tricky) that requires a discussion of the following points:

- whether s 2(3) is satisfied with respect to Helen and her cat;
- whether the Animals Act 1971 covers property damage;
- extent of Graham's liability for Helen's injury;
- extent of Graham's liability for Fiona's injury;
- any defences available to Graham; and
- other courses of action open to Helen and Fiona.

Answer

The Animals Act 1971 divides animals into dangerous and non-dangerous species. By s 6(2) of the 1971 Act, a dangerous species is one which is not commonly domesticated in the British Isles and whose fully grown animals normally have such characteristics that they are likely, unless restrained, to cause severe damage, and that any damage that they may cause is likely to be severe. Consequently, an Alsatian dog is a non-dangerous species. By s 2(2) of the Act, the keeper will be liable for the damage caused by an animal which does not belong to a dangerous species if:

(a) the damage is of a kind which the animal, unless restrained, was likely to cause or which, if caused by the animal, was likely to be severe; and

(b) the likelihood of the damage or of its being severe was due to characteristics of the animal which are not normally found in animals of the same species, or are not normally found except at particular times or in particular circumstances; and

(c) those characteristics were known to that keeper.

Under s 6(3), Graham is the keeper of the dog, as we are told that he is the owner. He will be *prima facie* liable if s 2(2) is satisfied. Considering first the position with Helen's cat, s 2(2)(a) is satisfied, as it is damage of the kind which the dog, unless restrained, is likely to cause. This wording is wide enough to cover damage by a dog running into the road or chasing a cat into the road.

Section 2(2)(b) is not so straightforward, because it requires the damage to be caused due to characteristics not normally so found in Alsatian dogs or only at particular times or in particular circumstances. In *Mirvahedy v Henley* (2004), the House of Lords held that s 2(2)(b) provided two separate bases of liability, namely that a keeper could be liable where the behaviour of the animal was not normally found in animals of the same species, and could also be liable where the behaviour, although not generally displayed by animals of that species, was normal in particular circumstances or at particular times. Thus, the House held that it was normal for horses to bolt when panicked by an external event. It would also seem normal for dogs to chase cats and so s 2(2)(b) is satisfied. Section 2(2)(c) is presumably satisfied as Graham would know of these tendencies. Hence Graham is liable for the damage to Helen's cat.[1]

Turning now to Helen, again s 2(2)(a) is satisfied, as the injury caused by a bite from an Alsatian is likely to be severe. Section 2(2)(b) is satisfied, because Alsatians are not normally vicious, except in the particular circumstances of being kept as guard dogs, and we are told that Graham keeps his dog to deter unwelcome visitors. Section 2(2)(c) is also satisfied, as Graham must know of the characteristics in his dog (*Cummings v Grainger* (1977)). Thus, Graham is liable to Helen for the bite, subject only to any defences contained within the Act. Sections 5(1) and 10, namely, that the harm was due wholly to Helen's fault or that Helen was contributorily negligent, do not apply on the facts we are given. Also, Graham cannot rely on s 5(3). Although s 5(3) exempts a keeper from liability for damage caused by an animal kept for protection of persons or property where keeping it for that purpose was not unreasonable, it only covers damage caused to trespassers, and Helen never entered Graham's property and so was never a trespasser. The *volenti* defence contained in s 5(2) is also clearly inapplicable. There may also have been a breach of s 1 of the Guard Dogs Act 1975, but s 5(1) of that Act expressly provides that breach shall not confer a civil right of action.

Turning now to Fiona, s 2(2)(a) is satisfied because, as we have argued earlier, the wording of s 2(2)(a) is wide enough to cover a dog running into the road. It is submitted that s 2(2)(b) is satisfied, as an Alsatian dog would not normally run into the road except in the particular circumstances of a guard dog chasing a perceived intruder from the premises it was guarding: *Mirvahedy*. Section 2(2)(c) is also satisfied because Graham knows of this characteristic. Thus, s 2(2) is satisfied in respect of Fiona and, as the harm was caused by the dog under s 2(2), there is no requirement of foreseeability. Consequently, Graham is liable, subject only to the defences in the Act. These have been considered with respect to Helen, and none could be relied on by Graham, who is liable for the injury suffered by Fiona.

Graham could argue that the harm to Fiona was caused not by the dog, but by Helen running into the road rather than along the pavement, so that Helen's action was a *novus actus interveniens* which broke the chain of causation. The act of a third party may break the chain of causation where it is something unwarrantable, a new cause which disturbs the sequence of events, something which can be described as either unreasonable or extraneous or extrinsic (*per* Lord Wright in *The Oropesa* (1943)). In Fiona's case, as the act of Helen was an involuntary one and not unreasonable, it will not break the chain of causation (*Scott v Shepherd* (1773)).

Helen and Fiona could also sue Graham in negligence for not taking reasonable steps to confine the dog within the limits of his property, and possibly in *Rylands v Fletcher* (1868), although whether liability exists in *Rylands* for the escape of an animal is debatable (*Read v Lyons* (1947)). However, *Rylands* has been held to cover the escape of caravan dwellers (*AG v Cooke* (1933)), so arguably it could cover animals. However, *Rylands* does not apply to personal injuries: *Transco v Stockport Metropolitan Borough Council* (2004).

Think point

1 One might also consider whether liability under the Animals Act 1971 extends to property damage. By s 11, damage is defined as including death or personal injury and property damage is not expressly covered. However, it has been argued in Rogers (ed), *Winfield and Jolowicz on Tort*, 16th edn, 2002, that as s 11 is not exhaustive, property damage is included, and it seems to have been allowed at common law (*Buckle v Holmes* (1926)).

Notes

Question 34

Henry owns a large dog which has a tendency to attack people in uniforms. Henry keeps the dog tied up in his garden with a substantial chain. Unfortunately, there is a latent defect in one link of the chain and, when Pat the postman goes to the front door of the house to deliver some letters, the dog attempts to attack Pat. The chain breaks and the dog bites Pat. Pat is taken to hospital and given an anti-tetanus injection, to which he suffers a rare and unforeseeable allergic reaction, and his leg has to be amputated. Richard, a policeman, calls to investigate the situation and the dog jumps over the garden fence and bites Richard. While Richard is doubled up in pain on the pavement, Steven, who Richard arrested for a drug offence a little while ago, sees Richard on the floor and kicks him in the head.

Advise Henry.

Answer plan

This question ranges over a number of aspects of liability for animals, both under the Animals Act 1971 and under other causes of action.

The following points need to be discussed:

- Henry's liability under s 2(2) to Pat;
- defences available to Henry in respect of Pat;
- Henry's liability under s 2(2) to Richard;
- defences available in respect of Richard; and
- Henry's liability for the action of Steven.

Answer

Under the statutory classification of the Animals Act 1971, Henry's dog is a non-dangerous species, because it is commonly domesticated in the British Isles (s 6(2)). By s 6(3), Henry is the keeper of the dog, as he is the owner. By s 2(2), the keeper of an animal belonging to a non-dangerous species is liable for the damage caused by the animal if:

(a) the damage is of a kind which the animal, unless restrained, was likely to cause or which, if caused by the animal, was likely to be severe; and
(b) the likelihood of the damage or of its being severe was due to characteristics of the animal which are not normally found in animals of the same species, or are not normally found except at particular times or in particular circumstances; and
(c) those characteristics were known to that keeper.

Considering now Henry's liability to Pat, s 2(2)(a) is satisfied, because the bite from a large dog is likely to be severe. The tendency to attack persons in uniform is not a characteristic of dogs (see, for example, *Kite v Napp* (1982)), so s 2(2)(b) is satisfied. Section 2(2)(c) is also satisfied because this characteristic would be known to Henry (see *Cummings v Grainger* (1977) and *Mirvahedy v Henley* (2003) for liability under s 2(2) generally). Thus, Henry is liable for the damage caused to Pat, subject only to the defences within the 1971 Act.

These defences include *volenti* (s 5(2)), contributory negligence (s 10) or that the damage was wholly due to the fault of the person suffering it (s 5(1)). Section 5(3) also provides a defence against trespassers, but this would not apply to Pat (see s 2(6) of the Occupiers' Liability Act 1957). It should be noted that neither the act of a stranger nor an act of God provide a defence to s 2(2), as they are not mentioned in the Act. Thus, the fact that the dog broke free from the chain due to a latent defect in the chain is not a defence, as liability under s 2(2) does not require negligence (*Curtis v Betts* (1990); *Mirvahedy v Henley* (2003)). Henry is liable for the bite suffered by Pat and he is also liable for the medical consequences of the anti-tetanus injection, because liability under the Act is strict and subject only to the defences contained within the Act. There is thus no necessity for the damage suffered to be reasonably foreseeable; it merely has to be a direct consequence of the action of the animal, that is, *Re Polemis* (1921) is the appropriate test of recovery of damage. In any event, even if the foreseeability was required, as Henry must take his victim as he finds him (*Dulieu v White* (1901)), that is, with an allergy to tetanus injections, or if the need for such an injection is foreseeable,

Henry will be liable for its consequences (*Robinson v Post Office* (1974)). Hence, Henry will be liable for both the bite and the loss of Pat's leg.

Turning now to Richard, following our discussion above, Henry will be liable to Richard for the bite and none of the statutory defences are valid. (Note that Richard is not a trespasser in this case, because he has not entered Henry's property.) The question arises as to whether Henry is liable for the kick perpetrated by Steven. Section 2(2) states that the keeper is liable for damage caused and, as we have seen, there is no requirement of foreseeability, merely directness. However, Henry could argue that the kick by Steven is a *novus actus interveniens* which breaks the chain of causation, that is, the injury from the kick was not caused by his dog and therefore that harm does not come within s 2(2). In *Re Polemis*, where directness was considered, Scrutton LJ stated that indirect damage meant damage caused by the 'operator of independent causes having no connection with the ... act, except that they could not avoid its results'. Where it is alleged that the act of a third party, over whom the defendant has no control, has broken the chain of causation, then it must be shown that the act was something unwarrantable, a new cause which disturbs the sequence of events. It must be something which can be described as either unreasonable or extraneous or extrinsic (*per* Lord Wright in *The Oropesa* (1943)). Thus, the defendant will remain liable if the act of the third party is not truly independent of the defendant's act. In *Knightley v Johns* (1982), the Court of Appeal held that negligent conduct was far more likely to break the chain of causation than non-negligent conduct, so it would follow that a deliberate act is even more likely to break the chain and be found to be truly independent of the defendant's original act.

In the circumstances, the act of Steven is unreasonable, extraneous, extrinsic and deliberate and would breach the chain of causation, so that Henry would not be liable for those consequences.

Henry could not be sued by Pat in negligence, as there has been no breach of duty on his part, as we are told that the chain was substantial, but had a latent defect. Henry could be sued in negligence by Richard as, once the dog broke free, Henry would have been negligent in not securing the dog if he was aware of the broken dog chain.

Richard could possibly sue Henry under *Rylands v Fletcher* (1868), although whether liability exists for the escape of an animal is debatable (*Read v Lyons* (1947)). However, the rule in *Rylands* has been held to be applicable in the case of an escape of caravan dwellers (*AG v Cooke* (1933)), so arguably it could cover the escape of animals. But even if *Rylands* did apply to the escape of an animal, it does not cover personal injuries: *Transco v Stockport Metropolitan Borough Council* (2004).

Notes

Question 35

Jenny, who lectures in zoology, has a pet South African monkey called Nigel. Nigel has been hand reared since he was born and is quite tame. One day, Nigel opened a window catch and climbed out of Jenny's house and went through an open window into the house of Jenny's neighbour, Angela. Angela's mother, Maria, was visiting at the time and, as Maria has a phobia about monkeys because she was bitten by one as a child, she panicked and ran through the glass back door, cutting herself extensively. She went to hospital by ambulance and, while she was at the hospital, a thief entered by the broken back door and stole some of Angela's property.

Advise Angela and Maria.

Answer plan

The question is a little different from the standard animals question, in that it involves a dangerous species, together with a consideration of the damage for which its keeper is liable.

The following points need to be discussed:

- definition of a dangerous species;
- liability for damage caused – extent and limitations; and
- other causes of action.

Answer

We first have to decide whether Nigel belongs to a dangerous or non-dangerous species. By s 6(2) of the Animals Act 1971, a dangerous species is a species:

(a) which is not commonly domesticated in the British Isles; and

(b) whose fully grown animals normally have such characteristics that they are likely, unless restrained, to cause severe damage or that any damage that they may cause is likely to be severe.

It should be noted that, by s 6(2), it is the species which must be dangerous and not the particular animal in question. Thus, the fact that Nigel is tame does not take him out of the category of dangerous species. In addition, s 6(2) requires that the animal be of a type which is not commonly domesticated in the British Isles – the fact that Nigel might belong to a species which is commonly domesticated in South Africa again will not take Nigel out of his classification. Thus, s 6(2)(a) is satisfied. Section 6(2)(b) is satisfied, as the bite from a fully grown monkey is likely to be severe. Thus, both heads of s 6(2) are satisfied and Nigel belongs to a dangerous species. By s 2(1), the keeper is liable for any damage caused, except where the Act provides a defence. There is no restriction on the damage caused by Nigel or the damage Nigel is likely to cause, or whether or not that damage is severe. It is also clear from the wording of the Act that there is no requirement that the damage be foreseeable; it is enough that it is caused by the animal. Jenny is the keeper of the animal under s 6(3), as we are told that Jenny owns Nigel.

From s 2(1), it follows that Jenny is liable for the harm caused by Nigel. The only defences available to Jenny are those contained within the Act, namely, *volenti* (s 5(2)), contributory negligence (s 10), the defence with regard to trespassers and guard dogs (s 5(3) and (1)), and where the damage is wholly due to the fault of the person suffering it. Clearly, s 5(2) and (3) is not relevant to Jenny, but could she claim that the injury suffered by Maria was wholly due to her fault in running through the glass door? As s 2(1) makes the keeper liable for the damage caused (subject to the statutory defences), s 5(1) covers the situation where the victim causes the damage wholly by himself. We should thus ask whether Maria's act of running through the door was a *novus actus interveniens* which broke the chain of causation, that is, that the appearance of Nigel merely provided the opportunity for Maria to be the author of her own misfortune. The problem for Jenny in running this defence is the well-established rule that a tortfeasor takes his victim as he finds him (*Dulieu v White* (1901)) and, in this case, the victim has a phobia about monkeys. No question of foreseeability arises under s 2(1) (though even if it did, it would be disposed of by the above rule: see *Robinson v Post Office* (1974); *Bradford v Robinson Rentals* (1967)). The act of the claimant may break the chain of causation where his act is so careless that his injury cannot be attributed to the fault of the defendant. Comparing *McKew v Holland and Hannen and Cubitts* (1969) with *Wieland v Cyril Lord Carpets* (1969), it seems clear that to constitute a *novus actus interveniens* on the part of the claimant, the act must be unreasonable. As Jenny must take Maria as she finds her, that is, with a phobia about monkeys, Maria's acts are not likely to be found so unreasonable as to constitute a *novus actus interveniens*. Again, it is settled law that if a person, in the agony of the moment, causes himself harm, the act causing the harm will not necessarily break the chain of causation: *Jones v Boyce* (1816). Hence, it is submitted that Maria's action will not constitute a *novus actus interveniens*, but that contributory negligence under s 10 would be a more appropriate defence (if any).

As regards the theft of property, the damage has been caused by a third party, so the question arises as to whether or not the act of the thief caused the damage rather than Nigel, that is, was the theft a *novus actus interveniens*? Where it is alleged that the act of a third party, over whom the defendant has no control, has broken the chain of causation, it must be shown that the act was something unwarrantable, a new cause which disturbs the sequence of events. It must be something which can be described as either unreasonable or extraneous or extrinsic (*per* Lord Wright in *The Oropesa* (1943)). Thus, the defendant will remain liable if the act of the third party is not truly independent of the defendant's act. In *Knightley v Johns* (1982), the Court of Appeal held that negligent conduct was far more likely to break the chain of causation than non-negligent conduct, so it would follow that a deliberate act is even more likely to break the chain and be found to be truly independent of the defendant's original act. The problem facing Jenny is that we are told that the thief entered by the broken back door, which suggests that the act of the thief may not be truly independent of Jenny's original act, in that the thief may not have entered the premises had the back door not been broken. If the court were to make such a finding, Jenny would be liable for the loss resulting from the theft. It is not likely that Jenny would succeed in claiming that the true cause of the theft was a *novus actus interveniens* by Maria in failing to secure the back door before going to hospital. Maria's actions seem reasonable in the 'agony of the moment' caused by Jenny's original tort, and would not break the chain of causation (*Jones v Boyce* (1861)). Although in *Stansbie v Troman* (1948), it was held that the act of a thief did not break the chain of causation, this was explained by Lord Goff in *Smith v Littlewoods Organisation* (1987) as being due to the contractual relationship between the parties in question.

Jenny could also be liable to Maria in negligence. There would be no difficulty in establishing a duty of care and breach of that duty, and the problem of causation (that is, did Maria herself cause her injuries?) has already been considered above. Similarly, with Angela, it would be straightforward enough to show the existence of a duty of care and breach of that duty (and again, we have considered the problem of causation above).

Jenny might also be liable under the rule in *Rylands v Fletcher* (1868) for the escape of Nigel if the rule applies to animals. This was doubted in *Read v Lyons* (1947), but the rule has been held to cover the escape of caravan dwellers (*AG v Cooke* (1933)), so by analogy it could cover the escape of an animal. However, even if *Rylands* does cover the escape of an animal, it does not cover personal injuries: *Transco v Stockport Metropolitan Borough Council* (2004).

Jenny could also be liable to Angela in nuisance. Although the escape of Nigel was an isolated event, in *Bolton v Stone* (1951), it was stated that although a nuisance could not arise from an isolated happening, it could arise from a state of affairs, albeit temporary. Thus, in *Midwood v Manchester Corp* (1905), a gas explosion was held to be a nuisance because, although it was an isolated event, it was due to a pre-existing state of affairs, namely, the build-up of gas. Hence, Angela could argue that the escape of Nigel was due to a wrongful state of affairs on Jenny's property, namely, that Nigel was not kept within Jenny's property. Whether Maria could sue in nuisance is a difficult point – it has been held traditionally that as the tort of nuisance protects interests in land, the claimant must have an interest in the land affected to sue (*Malone v Laskey* (1907)). Despite the relaxation of this requirement by the Court of Appeal in *Khorasandjian v Bush* (1993) and *Hunter v Canary Wharf Ltd* (1996), the House of Lords reinstated this requirement when it decided *Hunter* (1997). Thus, Maria lacks the necessary interest in land to sue in nuisance and, in addition, in *Hunter* (1997), the House of Lords held that personal injuries could not be recovered in an action in nuisance. Thus, Maria has no right in nuisance.

Notes

11 Defamation

Introduction

Questions on defamation appear regularly in examination papers. Defamation is a major topic and encompasses a considerable volume of law. In practice, examiners tend to concentrate on several specific topics, notably the defences of fair comment and qualified privilege, although students also will have to have a good grasp of the elements of liability and of the provisions of the Defamation Act 1996.

Checklist

Students must be familiar with the following areas:

(a) distinction between libel and slander;

(b) defamatory statements and innuendoes;

(c) reference to claimant;

(d) publication; and

(e) defences, with especial references to fair comment and qualified privilege.

Question 36

Alfred, a well-known and successful businessman, held a large party at his country house. Beryl, who once worked for Alfred in public relations but was dismissed, is now a reporter. She writes an article in the *Daily Globe*, in which she says: 'Alfred, who makes his money by rationalising companies, that is, by throwing people out of work, held a party at his house for the sycophants who work for him. Whether they would be so happy if they were aware of his bizarre view of business ethics during his recent takeover bid for Alpha plc is uncertain. Certainly, the investigation by the takeover panel will make "interesting reading".' The next day, as Alfred is walking into his office, Cedric, who was recently made redundant during Alfred's takeover of Alpha, sees him and shouts: 'You are a villain who thinks only of yourself. I hope they put you in jail for years over your takeover.'

Advise Alfred and his guests of the legal situation.

Answer plan

This is a typical defamation question – typical, in that it involves the elements of both liability and defences.

The following points need to be discussed:

- whether Beryl's statement is defamatory of Alfred;
- whether Beryl's statement is defamatory of the guests – problem of class defamation;
- defences available to Beryl, especially justification and fair comment;
- whether Cedric's statement is defamatory of Alfred; and
- defences available to Cedric.

Answer

We must consider whether Alfred and his guests have been defamed by Beryl and the *Daily Globe,* and whether Alfred has been defamed by Cedric.

The newspaper article by Beryl is in permanent form, and so any defamation will take the form of libel and will be actionable without any need to prove special damage. To succeed in an action for defamation, Alfred must prove that the statement complained of was defamatory, that it could reasonably be understood to refer to Alfred and that it was published to a third party.

The usual test for a statement being defamatory is that it tends to lower the claimant in the estimation of right-thinking members of society generally (*Sim v Stretch* (1936)) or exposes him to hatred, contempt or ridicule (*Parmiter v Coupland* (1840)). The statement regarding Alfred contains three possible defamatory elements, namely, the allegation that Alfred employs sycophants, that he has a bizarre view of business ethics and that he is being investigated by the takeover panel.

The first allegation may well be defamatory and Alfred could plead a false innuendo, that is, that the words contain a secondary meaning that he is incapable of choosing employees correctly, which would be defamatory of an eminent businessman. This would be a question for the jury to decide. In *Hartt v Newspaper Publishing* (1989), the Court of Appeal held that the approach to adopt was that of the hypothetical ordinary reader who was neither naive nor unduly suspicious, but who might read between the lines and be capable of loose thinking. This test was also applied by the Privy Council in *Bonnick v Morris* (2002). The statement that Alfred has a bizarre view of business ethics is defamatory (*Angel v Bushell and Co* (1968)), as it is suggesting a lack of honesty or probity.

The final part of Beryl's statement concerning the investigation by the takeover panel needs careful consideration. In *Lewis v Daily Telegraph* (1964), it was held by the House of Lords that to say a person was being investigated for fraud was not the same as saying that he was guilty of fraud. Thus, to say that Alfred is the subject of an investigation by the takeover panel is not, without more, defamatory.

It should perhaps be noted here that the test of the defamatory nature of a statement is its effect on right-thinking members of society. The fact that Beryl's statements might not cause Alfred's friends or business colleagues or employees to think any the less of him is not relevant (*Byrne v Deane* (1937)).

Next, we shall consider the guests: the allegation that they are sycophants is defamatory, as it would expose them to ridicule or contempt (*Parmiter v Coupland* (1840)).

The statement has clearly been published to a third party, but the problem for the guests is that we are told that it was a large party. Thus, the problem arises as to whether a group or class can sue when it has been defamed as an entity. In *Knupffer v London Express Newspapers* (1944), the House of Lords held that in class defamation, a member of the class could sue only if the words point particularly to the claimant or the class was so small that the words must necessarily refer to each member of it. Beryl's words do not particularly point to any guest, so whether the guests can sue on the statement will depend on the size of the class, that is, the number of guests. Unfortunately, we are given no indication of this in the facts of the question but, should the class be small enough, the necessary elements of the tort of defamation would be present for the guests.

Let us now consider any defences which are available to Beryl and the *Daily Globe*.

Considering the statement concerning Alfred, Beryl and the *Daily Globe* could rely on the defence of justification, that is, truth. This would be a valid defence for the allegation regarding the investigation by the takeover panel (assuming it is true). It would seem that it would be an extremely difficult defence to establish in respect of the allegation that Alfred employs sycophants and, as regards the business ethics allegation, difficulties of proof could arise for the defendants unless the takeover panel investigation substantiated these claims. Thus, their defence would be limited to the investigation allegations (if such an allegation were held to be defamatory, which is unlikely, as previously submitted).

The defendant may also raise the defence of fair comment, that is, that the statement is fair comment based on true facts made in good faith on a matter of public interest.

The courts define public interest widely (*London Artists v Littler* (1969)), and the activities of a prominent businessman would be a matter of public interest. But the comments must be based on true facts and, as we have seen, this truth may be difficult to establish for the comments concerning sycophants and business ethics. An additional problem arises in that the statements must be those of opinion and not of fact. It may be that a court would find the statement that Alfred employs sycophants to be a statement of opinion – see *Dakhyl v Labouchere* (1908) (though it must still be based on true facts: see *Merivale v Carson* (1887)) – but the statement regarding business ethics does appear to be more of a statement of fact. By 'fair', we mean that the defendant honestly believed the opinion expressed (*Slim v Daily Telegraph* (1968)) and not that a reasonable person would agree with the opinion (*Silkin v Beaverbrook Newspapers* (1958)). Although Beryl's comment may be fair in this respect, the defence can be rebutted by showing that the defendant acted out of malice (*Thomas v Bradbury Agnew* (1906)). The burden of proving malice will be on the claimant (*Telnikoff v Matusevich* (1991)), although the defendant will still have to show that the facts on which the comment was based were true and the comment was objectively fair, in that anyone, however prejudiced or obstinate, could honestly have held the views expressed. In view of the fact that Beryl was dismissed by Alfred, malice may be found on her part but, providing that the *Daily Globe* did not act maliciously, it will not be tainted with Beryl's malice (*Lyon v Daily Telegraph* (1943)).

Overall, therefore, it seems unlikely that either Beryl or the *Daily Globe* could rely on the defence of fair comment. In these circumstances, if Alfred believes that Beryl or the *Daily Globe*'s defence to his action has no realistic prospect of success, he could use the 'fast track' procedure provided by ss 8 and 9 of the Defamation Act 1996. Under this procedure, damages are assessed by a judge and not a jury, and are limited to £10,000. This procedure is only available where one side's case has no realistic prospect of success (s 8(2) of the 1996 Act). In addition to damages, Alfred may be able to obtain a declaration that the statement was false and defamatory, a published apology and an injunction to restrict further publication (s 9 of the 1996 Act). The *Daily Globe* could rely on an apology as a defence under the Libel Act 1843 if the statement was published with malice and without gross negligence, if an apology was published as soon as possible and a payment has been made into court by way of amends.

Turning to the guests and the allegation that they are sycophants (assuming that the guests can overcome the class problem), the only defence available to Beryl and the *Daily Globe* would appear to be fair comment. However, as we have seen from our discussion regarding Alfred, this defence is unlikely to succeed. The guests should also be advised of the possibility of using the fast track procedure as described above, as it would seem that the defendant's case has no realistic chance of success, thus satisfying s 8(2) of the Defamation Act 1996. The *Daily Globe* would also have available the apology defence under the Libel Act 1843.

In addition, both Alfred and his guests should be advised that a decision as to whether to institute proceedings should be made reasonably quickly. The limitation period for defamation proceedings is now one year (s 5 of the Defamation Act 1996), although this period may be extended under s 32A of the Limitation Act 1980 (as amended), but the court will require a satisfactory explanation for the delay (*Steedman v British Broadcasting Corp* (2001)).

Finally, we must consider Cedric's statement. This is in transient form and so it is slander and, normally, special damage would have to be shown for it to be actionable. However, where the words impute a crime punishable by imprisonment (*Hellwig v Mitchell* (1910)) or are calculated to disparage the claimant in any office, profession, calling, trade or business carried on by him (s 2 of the Defamation Act 1952), there is no need for the claimant to prove special damage. As Cedric's words fall into both categories, they are *prima facie* actionable. However, spoken words are not actionable where they amount to mere abuse or insult (*Parkins v Scott* (1862); *Lane v Holloway* (1968)). The test seems to be whether the statements would have been taken by a listener as those made in the heat of the moment, or whether they did contain a serious allegation.

In applying this test, it is submitted that no liability arises in respect of Cedric's statement.

Notes

Question 37

The Westfield Chamber of Commerce decides to set up a fund to allow a promising young businessman to spend some months in Europe studying European business methods. A committee consisting of Diana, Edward and Fenella is set up to consider applications. An application is received from George, and Diana circulates a memo to Edward, saying 'I understand that George is on the point of insolvency. He does not seem to be a suitable candidate'. Edward also circulates a memo, stating 'George is incompetent and not fit to represent Westfield in Europe'. Edward types this himself, but leaves a copy on the photocopying machine where it is seen by Henry. Edward's company recently tendered for some business with George's company, but failed to obtain the contract.

Advise George.

Answer plan

Again, a standard defamation question, requiring mostly a discussion of the defence of qualified privilege and fair comment.

The following points need to be discussed:

- whether Diana's statement is defamatory;
- whether Diana can claim qualified privilege or fair comment;
- effect of possible malice on Edward's defences; and
- possible evasion of qualified privilege defence by using negligent misstatement as a cause of action.

Answer

The statements by Diana and Edward are in permanent form, so any defamation that has occurred will take the form of libel and will be actionable without proof of any special damage.

In order to succeed in an action for defamation, George will have to prove that the relevant statement was defamatory, that it referred to him and was published to a third party.

The usual test for a statement being defamatory is that it tends to lower the claimant in the estimation of right-thinking members of society generally (*Sim v Stretch* (1936)) or exposes him to hatred, contempt or ridicule (*Parmiter v Coupland* (1840)). Diana's statement appears, at first sight, to meet this criterion: in *Read v Hudson* (1700), it was held to be defamatory to impute insolvency to a trader, even though there was no suggestion of discreditable conduct. Also, if the statement contains the false innuendo that George is not competent in his business or profession, that will clearly be defamatory (*Capital and Counties Bank v Henty* (1882)). The fact that George's friends or business colleagues might regard insolvency as something that might happen to even the most talented businessman is not relevant to the issue of whether the statement is defamatory: the statement must be judged by the standard of right-thinking members of society generally, not just the claimant's friends (*Byrne v Deane* (1937)).

Edward's statement is clearly defamatory, reflecting adversely on George's competence. Both Diana's and Edward's statements refer to George by name and Diana has published the name to a third party, namely, Edward and Fenella. Edward has published the name both to Diana and Fenella and also to Henry, as negligent publication to a third party is sufficient publication (*Theaker v Richardson* (1962)). This is assuming, of course, that Henry understands the defamatory nature of the statement and its reference to George (*Sadgrove v Hole* (1901)).

Prima facie, therefore, George can establish the elements of the tort of defamation against Diana and Edward. Next we shall consider any defences available.

Diana may be able to avail herself of the defence of justification, that is, truth, providing that George is in fact close to insolvency. If this is not the case, Diana may seek to rely on the defences of fair comment and qualified privilege. The first of these defences applies where the statement is fair comment based on true facts made in good faith on a matter of public interest. The courts define public interest widely (*London Artists v Littler* (1969)), and the award in question would be a matter of public interest. The comment must be based on true facts which are stated in the comment. The problem for Diana is that the comment that George is not a suitable candidate is based on a fact (that he is close to insolvency), but we do not know whether that fact is true.

By fair comment, we mean that Diana must have honestly believed the opinion (*Slim v Daily Telegraph* (1968)) and not that a reasonable person would agree with the opinion (*Silkin v Beaverbrook Newspapers* (1958)). Although Diana's comment may be fair in this respect, the defence can be rebutted by showing that the defendant acted out of malice (*Thomas v Bradbury Agnew* (1906)). The burden of proving malice will lie on the claimant (*Telnikoff v Matusevich* (1991)). However, the defendant will still have to show that the opinions were honestly expressed and based on accurately stated facts: *Branson v Bower* (2001).

Although Diana may well be able to establish that the comment was fair, as there seems to be no evidence of malice, she still has to overcome the hurdle of basing the comment on true facts.

It seems that Diana's best defence would be to rely on qualified privilege, namely, that she was under a duty to make the statement to Edward and Fenella and they were under a corresponding duty to receive it (*Watt v Longsden* (1930)). The requisite duty would exist in this case, and Diana could rely on this defence unless she was acting maliciously. By malice, it is meant that the defendant had no honest belief in the truth of her statement (*Horrocks v Lowe* (1975)), and there is no reason to impute malice to Diana.

In *Kearns v General Council of the Bar* (2003), the Court of Appeal stated that the common interest situation test was not always useful, and that it would be more helpful to distinguish between cases where the communicating parties were in an existing and established relationship and cases where no such relationship had been established and the communications were between strangers. Privilege would attach much more readily in the established relationship situations. If we apply this test, it seems that Diana, Edward and Fenella are in an established relationship and could not be described as strangers, hence qualified privilege should arise whatever test is used.

It would be possible for George to circumvent Diana's defence of qualified privilege by suing Diana in the tort of negligent misstatement. This route was allowed by the House of Lords in *Spring v Guardian Assurance* (1994), despite the argument that it effectively allowed the defence of qualified privilege to be side-stepped. As a duty of care was imposed in similar circumstances in *Spring*, George would only have to prove breach, that is, untruth of the statement, as causation and foreseeability of damage would appear to pose no problems.

Edward could rely on the defence of justification if his statement were true. As regards fair comment, Edward's problem is that his statement seems to be one of fact, rather than opinion, and there is no sub-stratum of fact as in *Kemsley v Foot* (1952). Edward could of course argue that his statement should be interpreted as 'George is incompetent and therefore not fit to represent Westfield in Europe', and thus is comment based on fact. He would then have to show that the statement 'George is incompetent' is a true fact, which would be difficult to prove. If it were a comment, it would have to be fair in the sense discussed above, but George might well be able to show malice on Edward's part which would destroy the defence.

If Edward were to seek to rely on qualified privilege, then, although he would be able, like Diana, to show the required reciprocal duty or existing relationship regarding Diana and Fenella, this defence too can be destroyed by showing that Edward was actuated by malice. In any event, Edward could not rely on qualified privilege as regards the publishing to Henry, as Edward is under no duty to make the statement to Henry and Henry is under no duty to receive it (*Watt*).

It would seem, then, that George has a good case against Edward, but that Diana may be able to rely on qualified privilege as a defence to defamation, although she still has a problem as regards negligent misstatement.

In view of the strength of George's case against Edward, he should be advised that if it is decided that Edward's defence has no realistic prospect of success, George could avail himself of the 'fast track' procedure provided by ss 8 and 9 of the Defamation Act 1996. Under this procedure, damages are assessed by a judge and not a jury, and are limited to £10,000. In addition, George may be able to obtain a declaration that the statement was false and defamatory, a published apology and an injunction to restrict further publication (s 9 of the Defamation Act 1996).

George should also be advised that a decision as to whether to institute proceedings in defamation should be made reasonably quickly, as the limitation period for defamation actions is now one year (s 5 of the Defamation Act 1996). (This period may be extended under s 32A of the Limitation Act 1980 (as amended) but the court will require a satisfactory explanation for the delay (*Steedman v British Broadcasting Corp* (2001)).)

Question 38

Ian is a sports commentator for Eastland TV. He decides to make a programme on Eastleigh Rovers, a local amateur football team that has reached a regional cup final. In the programme, there is a shot of the team in a public house with the comment from Ian, 'This is how the team prepares on Friday night for its cup final match on Saturday'. In fact, the scene was shot on a Saturday night after a previous game. This film also shows John, the centre-forward, eating a hamburger with the comment from Ian, 'As a bachelor, John has to do his own cooking so he eats out a lot'. John is in fact married to Jane, who is most upset at this comment.

Eastleigh Rovers lose their cup final and Ian, in his post-match summary, states, 'They played appallingly badly, even by the standards of an amateur team'. The *Eastland Gazette* reviews the programme and match, repeats Ian's comments regarding the team playing badly and wonders whether this was due to John's poor diet.

Advise John, Jane and Eastleigh Rovers of any action they might have in defamation.

Answer plan

This is a wide-ranging question which covers the areas of innuendo, references to the claimant and re-publication.

The following points need to be discussed:

- commentary and slander – the Defamation Act 1952;
- Jane's ability to sue, despite not being expressly referred to;
- Eastleigh Rovers and class defamation; and
- the liability of Eastland TV for repetition of a defamatory statement.

Answer

The statements made by Ian in the TV programme are deemed to be publication in a permanent form by s 1 of the Defamation Act 1952. They may thus constitute libel and be actionable without proof of special damage.

For any of the potential claimants to sue in defamation, they must show that the statement complained of was defamatory, that it referred to them and that it was published to a third party.

Considering first John, the statement that he is a bachelor is not *prima facie* defamatory. However, when coupled with the true innuendo that John is married to Jane, the statement that he is a bachelor might lead people who know that he lives with Jane to assume that they are not in fact married (*Cassidy v Daily Mirror* (1929)). Thus, the statement is defamatory, as it would tend to lower John in the estimation of right-thinking members of society generally (*Sim v Stretch* (1936)) or expose him to hatred, contempt or ridicule (*Parmiter v Coupland* (1840)). The fact that John's friends might not think any the less of him for living with a woman to whom he is not married is not relevant, as the standard is that of right-thinking members of society (*Byrne v Deane* (1937)). It could be argued by Ian that the standards of right-thinking members of society alter with time, so that, for example, it is no longer defamatory to call a person a German, as in *Slazengers v Gibbs* (1916), or a Czech, as in *Linklater v Daily Telegraph* (1964). Given this, Ian could argue that to say that an adult male lives with a woman is no longer defamatory. As, however, it is still defamatory to make this allegation of a woman (s 1 of the Slander of Women Act 1891), it seems most illogical that it would not also be defamatory of a man.

Turning to Jane, Ian's statement concerning John obviously also carries the suggestion that Jane is living with John without being married to him: *Cassidy v Daily Mirror*. This is defamatory, and the fact that Jane is not referred to by Ian is no bar to her suing (*Morgan v Odhams Press* (1971)). The fact that Ian is innocent in this matter (for example, because he was mistaken or was even told that John was unmarried) is of itself no defence, as defamation depends on the fact of defamation, not the intent of the defamer (*Hulton v Jones* (1910)). Jane could also argue that the film of John eating out, plus the commentary suggesting that she does not do any cooking for John, is also defamatory, using the test in *Byrne*.

In both John's and Jane's case, there would be no problem in showing the statement referred to them and had been published to a third party.

The next question is whether the team, Eastleigh Rovers, can sue in defamation. The statement that the team prepares for a cup final by drinking the night before and the statement concerning how badly they played are both *prima facie* defamatory, and these statements were published to third parties. However, the statements concerning the team are an example of class or group defamation. In *Knupffer v London Express Newspapers* (1944), it was held by the House of Lords that in class defamation, a member of the class could not sue, unless the words pointed particularly to the claimant or that the class was so small that the words must necessarily refer to each member of it. It is submitted that a football team is such a small class that the individual member can sue.

Having established Ian's (and Eastland TV's) liability to these statements, we need to consider whether Ian can raise any successful defences. In respect of John and Jane, no common law defences seem available. However, both Ian and the TV company could make use of the offer to make amends defence contained within ss 2–4 of the Defamation Act 1996. By s 2(4), such an offer must be to make and publish a suitable correction and apology and to pay compensation. By s 3 of the 1996 Act, if such an offer is accepted, any defamation is ended and, if the parties cannot agree on compensation, this amount may be decided by the court.

If such an offer is not accepted by the claimant, the making of the offer is a valid defence. However, the defence is not available if the defendant knew that the statement could refer to the claimant and was both false and defamatory. The burden of proving this lies on the claimant. In *Milne v Express Newspapers Ltd* (2003), it was held that the amends defence would only fail where the defendant had actual knowledge of the facts and that actual knowledge would have provided reasonable grounds to believe that the words were false and defamatory. On the facts given, it seems that Ian and the TV company could avail themselves of this defence.

Considering the statement made about the team, there seems to be no defence to the allegations regarding drinking. As regards the allegation that they played appallingly badly, the defences available are justification, that is, truth, and fair comment. To establish the defence of fair comment, it will have to be shown that the statement was fair comment based on true facts made in good faith on a matter of public interest. The courts interpret public interest widely (*London Artists v Littler* (1969)), and a televised football match would certainly come under this heading. The comment must be one of opinion and not of fact, which is the case here. The comment must also be based on true facts, which must either be stated in the comment or be capable of being inferred from the comment (*Kemsley v Foot* (1952)). In a case such as the present, the comment is an opinion based on the fact in Ian's commentary – Ian does not have to set these all out again in detail before he gives his opinion (*McQuire v Western Morning News* (1903)). 'Fair comment' means that Ian must have honestly believed the opinion expressed (*Slim v Daily Telegraph* (1968)) and not that a reasonable person would agree with the opinion (*Silkin v Beaverbrook Newspapers* (1958)). Although Ian's comment may be fair in this respect, the defence can be rebutted by showing that the defendant acted out of malice (*Thomas v Bradbury Agnew* (1906)). The burden of proving malice will be on the claimant (*Telnikoff v Matusevich* (1991)). However, the defendant will still have to show that the facts on which the comment was based were true and the comment was objectively fair, in that anyone, however prejudiced or obstinate, could honestly have held the views expressed. Hence, overall it would seem that a defence of fair comment would be likely to succeed in the post-match comments.

The review in the *Eastland Gazette* constitutes a re-publication of the comments regarding the team and, by implication, re-publishes the statement regarding John and Jane. The question is whether any liability for this re-publication attaches to Eastland TV, or whether liability is solely that of the *Eastland Gazette*. In *McManus v Beckham* (2002), the Court of Appeal held that in re-publishing situations, the question was whether it was just that the defendant should be held liable for the damage. The court held that the test to apply was essentially one of reasonable foreseeability, but as this test might be too easily satisfied, that the court should ask whether the defendant actually either knew what he or she said was likely to be reported and repeated or whether a reasonable person in the position of the defendant would have known this.

Applying this test, the TV company could also be liable for the subsequent repetition of the story of their allegations in the *Eastland Gazette*.

Notes

Question 39

To what extent do you think that the law of defamation represents an unwarranted restriction on freedom of speech, particularly in the area of political comment?

Answer plan

This essay calls for a discussion of the elements of liability for defamation, together with those defences which are relevant to preserving freedom of speech.

The following points need to be considered:

- elements of liability;
- position with local authorities, political parties;
- relevant defences – consent, justification, absolute privilege, qualified privilege, fair comment; and
- defences under ss 1 and 2(4) of the Defamation Act 1996.

Answer

To consider whether the law on defamation represents any restriction on free speech, we must consider what constitutes defamation and what defences to defamation exist in law.

Let us start by looking at those persons who can sue in defamation. The basic rule is that only living persons can sue, so no restrictions exist at all on freedom of speech as regards dead persons. However, in law, a company is a person and can sue for defamatory statements affecting its business (*Metropolitan Saloon Omnibus Co v Hawkins* (1859)). But it has been held by the House of Lords in *Derbyshire County Council v Times Newspapers* (1993) (overruling *Bognor Regis Urban District Council v Campion* (1972)) that a local authority cannot sue for libel as regards its governing reputation. In view of the question being answered, it is interesting to note that the House of Lords decided that to hold otherwise would impose a substantial and unjustifiable restriction on freedom of speech. Similarly, in *Goldsmith v Bhoyrul* (1997), the High Court held that a political party could not sue in libel, as it would be an unjustified restriction on freedom of speech.

We next need to consider what constitutes defamation. The standard test is that proposed by Lord Atkin in *Sim v Stretch* (1936), namely, that a statement is defamatory if the words are 'words which tend to lower the plaintiff in the estimation of right-thinking members of society generally'. In the recent case of *Berkoff v Burchill* (1996), where the plaintiff was described as 'hideously ugly', it was held by the Court of Appeal that although insults which did not diminish a person's standing were not defamatory, a statement could be defamatory if it held up the plaintiff to contempt, scorn or ridicule or tended to exclude him from society, even if the statement did not impute disgraceful conduct or any lack of business or professional skill. This decision would appear to have made some inroads into freedom of speech, especially when one considers that it would be difficult to raise a defence against such an action. In *Hartt v Newspaper Publishing plc* (1989), the Court of Appeal held that in determining the meaning of the words, the approach adopted should be that of the hypothetical reader who was neither naive nor unduly suspicious, but who might read between the lines and be capable of loose thinking. This test was also adopted by the Privy Council in *Bonnick v Morris* (2002). The effect of this test is to make a great many statements potentially actionable, but it should be remembered that the standard is the objective one of the right-thinking member of society. Another factor which tends to widen the possible scope of liability is that there is no need for the claimant to be referred to by name – it is sufficient that the statement could be understood to refer to him (*Cassidy v Daily Mirror* (1929); *Morgan v Odhams Press* (1971)). In addition, the maker of a statement may be held liable for the re-publication of that statement where it is just that the defendant should be held liable for the damage. In *McManus v Beckham* (2002), the Court of Appeal held that the test to be applied is essentially one of reasonable foreseeability, but as this test might be too easily satisfied, that the court should ask whether the defendant actually either knew what he said was likely to be reported and repeated, or whether a reasonable person in the position of the defendant would have known this. It can thus be seen that defamation is a tort of potentially very wide scope, so we must turn to the defences that will limit liability and preserve freedom of speech.

Consent is a defence to defamation. For example, a person may consent to the publication of what would otherwise be defamatory material, for example, the 'My Wicked Life' type of newspaper interview. Another defence which probably prevents the bringing of a number of libel actions is that of justification or truth. It is sufficient in this respect to show that the substance of the allegation is true – it is not necessary to show that the statement is true in each and every particular. If more than one allegation is made against the claimant, then, by s 5 of the Defamation Act 1952, the defence will not fail merely because one of the allegations is untrue if that allegation does not materially affect the claimant's reputation having regard to the true allegation(s). Statements made on certain occasions carry absolute privilege, that is, no liability will attach to them, no matter how false or malicious they might be. Such occasions include statements made in Parliament, in judicial proceedings and in official communications. It can thus be seen that there are virtually no restrictions on freedom of speech on such occasions. There are, additionally, a number of situations to which qualified privilege attaches. This defence can be destroyed by showing that the defendant was actuated by malice, that is, that the defendant had no honest belief in the truth of his statement (*Horrocks v Lowe* (1975)). Perhaps the most important situation to which this defence attaches are the statements made by A to B concerning C, where both A and B have an interest in the statement (*Watt v Longsden* (1930)). The absence of the requisite interest on the part of either party is fatal to this defence (*Watt*). This is a defence that could be relevant in a variety of situations and, from the point of view of preserving freedom of speech, it should be noted that malice cannot be inferred merely because the maker of the statement is unreasonable or prejudiced or unfair (*Horrocks*). Although the defence of qualified privilege applies in references, in *Spring v Guardian Assurance* (1994), the House of Lords allowed a plaintiff to bring an action as regards an allegedly negligent reference via the tort of negligent misstatement. Had the action been brought in defamation, the plaintiff would have been met with the defence of qualified privilege, and would have had to show that the defendant acted out of malice. In *Spring*, the defence raised the point that the side-stepping of qualified privilege represented a restriction on freedom of speech, but the House of Lords held that, despite this, it was fair, just and reasonable to impose a duty of care in the circumstances. It could be argued that the effect of *Spring* is to impose a restriction on freedom of speech which would not have existed in the law of defamation. The House of Lords has recently considered the extent of the defence of qualified privilege in a case involving political comment: *Reynolds v Times Newspapers Ltd* (1999). It has been suggested by many commentators that the scope of the qualified privilege defence is too narrow and that a wide public figure defence should be created, whereby a public figure cannot sue in libel unless it is proved that the defendant was actuated by malice. This approach was rejected by the House of Lords, who upheld the traditional tests of duty and interest. Lord Nicholls stated the essential question that had to be answered was whether the public was entitled to know the information. In considering whether the allegations made attracted qualified privilege, a number of matters should be considered, including: the seriousness of the allegation; the extent to which the subject matter is of public concern; the source of the information; any steps taken to verify the information; and the claimant's comments. Thus, the test in *Reynolds* appears to be whether the defendant took reasonable care in establishing the truth of the story. The House of Lords went on to hold that the court should have particular regard to freedom of expression and be slow to conclude that a publication was not in the public interest, especially where the publication concerned matters of a political nature, and that any lingering doubts should be resolved in favour of publication.

The effect of *Reynolds* is that political debate in newspapers should be free, providing that journalists are responsible. Indeed, in *Loutchansky v Times Newspapers Ltd* (2001), the Court of Appeal held that in deciding whether there was a duty to publish defamatory words to the world at large, the standard to be applied was that of responsible journalism. Further, in *Bonnick v Morris* (2002), the Privy Council held that where a statement had a possible defamatory meaning that was not necessarily obvious to an ordinary, reasonable reader, the journalist could still rely on *Reynolds* to establish qualified privilege providing that the journalist had been responsible in reporting matters of public concern. There had been suggestions that publication to the public at large of allegations concerning public figures might fail to attract qualified privilege, as publication to the public was too wide a publication. *Reynolds* has disposed of these fears and protects the reasonable publication of political comment.

Another widely used defence in defamation actions is that of fair comment based on true facts made in good faith on a matter of public interest. The courts tend to define public interest very widely (*London Artists v Littler* (1969)) and, as in justification, it is the sting of the allegation that has to be true rather than each and every allegation. However, the statement must be comment, that is, it must be opinion rather than a factual statement, and distinguishing between opinion and fact can sometimes be difficult. Finally, the comment must be fair, which means that the defendant must have honestly believed the opinion (*Slim v Daily Telegraph* (1968)).

Because of the strictness of the common law rules regarding reference to the claimant and the relevance of extraneous matters which may not be known to the maker of the statement, ss 2–4 of the Defamation Act 1996 provides a defence in what might be called unintentional defamation. This defence involves an offer to make amends, which is an offer to make and publish a suitable correction and apology and to pay compensation. By s 3 of the 1996 Act, if such an offer is accepted, proceedings cease and, if the parties cannot agree on compensation, an amount may be set by the court. If an offer to make amends is not accepted, the offer is a valid defence to defamation proceedings: s 4 of the 1996 Act. However, the defence is not available if the defendant knew that the statement could refer to the claimant and was both false and defamatory. The burden of proving this lies on the claimant. Section 2 thus covers the situation where the defendant was unaware that he was referring to the claimant or was unaware that the material published was defamatory. It was held in *Milne v Express Newspapers Ltd* (2003) that this defence would only fail where the defendant had actual knowledge of the facts and that actual knowledge would have provided reasonable grounds to believe that the words were false and defamatory. In particular, negligence is insufficient to defeat the defence. Given the objective requirement of actual knowledge and subjective requirement of reasonable belief, the defence should succeed where the defendant has acted responsibly and provides a valuable safeguard for freedom of speech.

Section 1 of the 1996 Act also provides a defence for innocent dissemination (for example, by booksellers or newspaper vendors). Section 1 provides that a person has a defence if he can show that: (i) he was not the author, editor or publisher of the statement; (ii) he took reasonable care in its publication; and (iii) he did not know, and had no reason to believe, that what he did caused or contributed to the publication. Thus, the defence failed in *Godfrey v Demon Internet Ltd* (1998) as, although the defendant was not the 'author, editor or publisher' of the publication, the defendant failed to remove the defamatory material when he became aware of its defamatory nature.

In the law of defamation, a balance must be struck between protecting the reputation of persons and infringing freedom of speech (*Derbyshire County Council v Times Newspapers* (1993)). Liability in defamation is wide, but a number of defences are available which have the effect of protecting free speech. Given the variety and scope of these defences, and especially the width of qualified privilege since *Reynolds*, it is difficult to claim that the restrictions imposed by the law of defamation are unwarrantable, especially in the area of political comment. It could perhaps be argued that as public funding is not available for defamation actions, for very many people the law of defamation is irrelevant in view of the excessively high costs of proceeding with an action, and thus in practice, defamation represents a minimal restriction on freedom of speech. However, in *Joyce v Sengupta* (1993), a plaintiff was allowed to proceed in malicious falsehood as regards an alleged defamatory statement, and public funding is available for this tort. Thus, in those situations where malicious falsehood is a possible cause of action, public funding is available and the issue of cost becomes somewhat less important. Thus, the practical (or cynical) view that no real restriction on freedom of speech exists as so few people can afford to exercise their rights is perhaps no longer valid. Also, it should be noted that the Court of Appeal has recently held that aggravated damages may be awarded in malicious falsehood (*Khodaparast v Shad* (2000)). This may bring damages for this tort closer to those awarded by juries in defamation cases.

Notes

12 Trespass to the Person, to Land and to Goods

Introduction

Trespass is an area which may be tested by the examiner either in its own right or as part of a question, mostly involving, for example, occupiers' liability or nuisance.

There have, however, been a number of recent developments in the law of trespass, such as hostile touching, trespass to air space and false imprisonment of prisoners, which may jog the examiner's mind on the topic of trespass.

Checklist

Students must be familiar with the following areas:

(a) definition and elements of, and defences to, assault;

(b) definition and elements of, and defences to, false imprisonment;

(c) the rule in *Wilkinson v Downton* (1987);

(d) definition and elements of, and defences to, trespass to land, and especially trespass to airspace; and

(e) definition and elements of, and defences to, trespass to goods, and in particular title to lost goods and the allowance for improvement of goods.

Question 40

Javid, who was conducting a market survey, entered Keith's property in order to ask him some questions. Keith came to the door and said to Javid, 'If you have come to try to sell me anything, you can clear off', and raised his fist to Javid. This frightened Javid, who ran away, but tripped over and broke his leg. Keith immediately ran to help Javid. While he was bending over Javid and trying to help him, Lionel came along, assumed that Keith had hit Javid and took Keith to a police station, where he said, 'This man has hit an innocent man'. Keith was kept in custody in a very damp cell while inquiries were made, and was later released.

Advise Javid and Keith.

Answer plan

This question covers assault and false imprisonment. The test for recovery of damages in this tort needs to be discussed, together with the relevant provisions of the Police and Criminal Evidence Act 1984.

The following points need to be discussed:

- status of Javid – whether a visitor or trespasser;
- assault by Keith;
- Keith's liability for Javid's fall;
- false imprisonment by Lionel;
- defamation by Lionel; and
- false imprisonment by the police.

Answer

We should first consider the legal status of Javid, that is, whether when Javid entered Keith's property he was a visitor or a trespasser. In *Robson v Hallett* (1967), it was held that when a person enters premises for the purpose of communicating with the occupier, he is treated as having the occupier's implied permission to be there, until the visitor knows or ought to know that this permission has been revoked. Once this permission has been revoked, the visitor has a reasonable time to leave before he becomes a trespasser. Thus, Javid is a visitor, as Keith's permission has not been revoked. Even if Javid were to be a trespasser, although reasonable force may be used to eject him (*Green v Goddard* (1702)), this can only be done after he has been asked to leave the premises and been allowed a reasonable opportunity to do so (*Robson v Hallett* (1967)). Neither of these requirements have been met in Javid's case.

When Keith raises his fist to Javid, this is an assault. An assault is an attempt or threat to apply force to a person whereby that person is put in fear of immediate physical contact. In *Thomas v NUM (South Wales Area)* (1985), it was held by the High Court that where the plaintiff has no reasonable belief that the defendant can effect his purpose, there is no assault, even if the conduct is frightening. However, in Javid's case, even if Javid is by nature exceptionally timid, it would seem that he has a reasonable belief that Keith can effect his purpose. Hence, Keith has committed an assault on Javid, so we must consider whether Keith is liable for Javid's broken leg. The rule for remoteness of damage in trespass to the person is that the defendant is liable for all the direct consequences of the trespass (*Nash v Sheen* (1953)), that is, the test in *Re Polemis* (1921). Thus, Keith will be liable for Javid's broken leg. It seems unlikely that Keith could claim that Javid's carelessness amounted to a *novus actus interveniens*. In *Re Polemis*, it was held that the defendant was liable for all the damage directly resulting from his act. Keith would not be liable for any damage indirectly resulting, that is, damage due to the 'operation of independent causes having no connection with the act, except they could not avoid its result' (*per* Scrutton LJ). This gives rise to the possibility that he could claim that Javid's carelessness in tripping was a *novus actus interveniens* which broke the chain of causation. A comparison of *McKew v Holland and Hannen and Cubitts* (1969) with *Wieland v Cyril Lord Carpets* (1969) suggests that a subsequent act of the claimant will only be treated as breaking the chain of causation where the act is unreasonable, and it is not unreasonable for a person subjected to an assault to run away and concentrate on escaping rather than anything else. Thus, it would be more realistic of Keith to allege contributory negligence on Javid's part. This defence has been held to be applicable to battery (*Barnes v Nayer* (1986)), and the reasoning of the Court of Appeal would also seem to be appropriate to assault.

When Lionel takes Keith to the police station, he commits *prima facie* false imprisonment and possibly battery, as Lionel presumably restrains Keith in some way. False imprisonment is the total deprivation of the freedom of a person for any period, however short, without lawful justification (see *R v Bournewood Community and Mental Health NHS Trust* (1998) for a restatement by the House of Lords of the elements of false imprisonment). The only justification Lionel could claim is under the provisions of the Police and Criminal Evidence Act (PACE) 1984. By s 24(4), anyone, that is, a private citizen or a police officer, may arrest without warrant anyone who is in the act of committing an arrestable offence or anyone who he has reasonable grounds for suspecting of committing such an offence. However, s 24(4) does not apply where that offence has been committed. Under s 24(5), where an arrestable offence has been committed, any person may arrest without warrant anyone who is guilty of the offence or anyone who he has reasonable grounds to suspect to be guilty of it. This means of course that where an offence has been committed, there is a defence to false imprisonment, even if the person arresting arrests an innocent person. If an offence has not been committed, s 24(6) allows a police officer to arrest without warrant anyone he reasonably suspects to have committed an offence. Thus, PACE 1984 preserves the trap in *Walters v Smith* (1914), where it is necessary for a private person to prove that the offence in question has been committed by someone. The case of *R v Self* (1992) is a more recent example of the trap in action.

Lionel's problem is that as no arrestable offence has in fact been committed, he cannot invoke s 24(5) and he falls into the trap of *Walters* and will be liable for false imprisonment.

Lionel's statement that Keith has hit Javid is defamatory, refers to Keith and has been published to a third party, that is, heard by the police. This contains all the necessary elements of defamation, but will be covered by qualified privilege, as he is under a moral duty to make the statement and the police officer is under a legal duty to receive it (*Watt v Longsden* (1930)).

Recently, the Court of Appeal in *Kearns v General Council of the Bar* (2002) held that the duty-interest test was not always useful, and that it would be more helpful to distinguish between cases where the communicating parties were in an existing and established relationship and cases where no such relationship had been established and the communication was between strangers. The court held that privilege would attach much more readily in the established relationship situations. It is submitted, however, that *Kearns* was not intended to do away with qualified privilege in the well-established situation of a member of the public giving information regarding an alleged offence to a police officer.

When Keith was kept in custody, this was a lawful imprisonment under s 24(6) of PACE 1984 and, even if the conditions of imprisonment become or are intolerable, that does not render the detention unlawful (see *R v Deputy Governor of Parkhurst Prison ex p Hague: Weldon v Home Office* (1991), where the House of Lords expressly disapproved of *dicta* to the contrary in *Middleweek v Chief Constable of Merseyside* (1990)). However, in the *Hague* case, the House of Lords did state that such detention might give rise to a remedy in public law and, if the prisoner suffered an injury to his health, a remedy might lie in negligence. So, Keith cannot sue the police for false imprisonment, though if he had suffered any injury due to the dampness of the cell, he could sue in negligence. Keith could not sue Lionel for false imprisonment while Keith was in police custody, as Lionel merely gave information to the police, who effected Keith's arrest and detention (*Davidson v Chief Constable of North Wales* (1994)).

Notes

Question 41

Martin owns a house with a very large garden. Neil is taking his dog for a walk along the road bordering Martin's house, when the dog jumps over Martin's fence and runs into his garden to eat some flowers. Neil enters the garden to retrieve his dog who by now has run into Martin's greenhouse. While Neil is in the greenhouse, Martin sees him and shuts the door, saying, 'stop there, you thief, I am phoning the police'. Neil, who knows that he can explain his presence to the police, is quite happy to stay in Martin's greenhouse admiring Martin's collection of exotic plants. The police arrive in a few minutes, and no charges are brought against Neil. Neil leaves his jacket in Martin's greenhouse, but Martin refuses to return it to Neil until Neil pays Martin compensation for damage Martin claims was done to the plants in his garden by Neil.

Advise Martin and Neil of any legal consequences of their actions.

Answer plan

This question covers trespass to land and false imprisonment. The question as to whether trespass to land can be committed negligently must be discussed, together with the elements and defences to false imprisonment.

Thus, the answer should consider:

- Neil's trespass via the dog;
- Neil's trespass on Martin's property;
- false imprisonment by Martin, and the provisions of the Police and Criminal Evidence Act (PACE) 1984 and common law defences;
- distress damage feasant; and
- other causes of action, for example, negligence, nuisance, the Animals Act 1971 and defamation.

Answer

When Neil's dog enters Martin's property, Neil has committed trespass to land through his dog and it seems, from *League Against Cruel Sports v Scott* (1985), that trespass to land through animals can be committed negligently. When Neil enters Martin's land to retrieve the dog, he too has committed trespass to land, as clearly Neil intended to enter upon Martin's land, which is sufficient – there is no need to show that Neil intended to trespass (*Conway v Wimpey* (1951)).

When Martin shuts Neil in the greenhouse, Martin has committed the tort of false imprisonment, which consists of the total deprivation of the freedom of a person for any period, however short, without lawful justification (see *R v Bournewood Community and Mental Health NHS Trust* (1998) for a recent restatement by the House of Lords of the elements of false imprisonment). The fact that Neil is quite happy to remain in the greenhouse is not relevant to liability, though it would be relevant to any issue of damages, should this arise. Two defences are relevant to Martin's actions here, namely, the provisions of PACE 1984 and the common law defences. By s 24(4), anyone, that is, a private citizen or a police officer, may arrest without warrant anyone who is in the act of committing an arrestable offence or anyone who he has reasonable grounds for suspecting of committing such an offence. However, s 24(4) does not apply where the offence has been committed. Under s 24(5), where an arrestable offence has been committed, any person may arrest without warrant anyone who is guilty of the offence or anyone who he has reasonable grounds to suspect to be guilty of it. This means, of course, that where an offence has been committed, there is a defence to false imprisonment, even if the person arresting arrests an innocent person. If an offence has not been committed, s 24(6) allows a police officer to arrest without warrant anyone he reasonably suspects to have committed an offence. Thus, PACE 1984 preserves the trap in *Walters v Smith* (1914), where it is necessary for a private person to prove that the offence in question has been committed by someone. The recent case of *R v Self* (1992) is an example of the trap in action.

However, as Neil is a trespasser, for he has entered Martin's land without invitation and his presence is objected to (*Addie v Dumbreck* (1929)), Martin may use a reasonable degree of force to control his movement (*Harrison v Rutland* (1893); *Alderson v Booth* (1969)).

As regards Martin's retention of Neil's jacket, this is a conversion of Neil's goods, as Martin is performing a positive wrongful act or dealing with the goods in a manner which is inconsistent with the rights of the owner (*Maynegrain v Campafina Bank* (1984)). Neil can therefore sue Martin in conversion and, by s 3(2) of the Torts (Interference with Goods) Act 1977, the remedies available are an order to deliver the goods to Neil or to pay damages or an order to deliver, with the alternative of paying damages. The defence of distress damage feasant makes it lawful for an occupier of land to seize any chattels which are unlawfully on his land and have done damage therein, and to detain them until payment of compensation for the damage. The problem for Martin is that Neil's jacket has not caused actual damage and thus his jacket cannot be lawfully detained (*R v Howson* (1966)).

A number of other causes of action are disclosed by the facts of the problem. When Martin says 'Stop there, you thief, I am phoning the police', this is a defamatory statement. Although it is slander, as it imputes a crime punishable by imprisonment, it is actionable without proof of special damage (*Hellwig v Mitchell* (1910)). It refers to Neil, but the question arises as to whether it has been published to a third party. If any person heard Martin's statement, Neil can sue Martin in defamation, but if no one other than Martin or Neil heard the statement, there is no publication. When Martin alleged to the police that Neil was a thief, publication would be covered by qualified privilege, as Martin is under a moral duty to make the statement and the police are under a legal duty to receive it (*Watt v Longsden* (1930)).

Recently, the Court of Appeal in *Kearns v General Council of the Bar* (2002) held that the duty-interest test was not always useful, and that it would be more helpful to distinguish between cases where the communicating parties were in an existing and established relationship and cases where no such relationship had been established and the communication was between strangers. The court held that privilege would attach much more readily in the established relationship situations. It is submitted, however, that *Kearns* was not intended to do away with qualified privilege in the well-established situations of a member of the public giving information regarding an alleged offence to a police officer.

Martin could also sue Neil in nuisance, as there has been an unreasonable interference with Martin's use or enjoyment of his land and, although this was an isolated event, it was due to a wrongful state of affairs, that is, the dog not being on a lead or not being properly controlled (*Pitcher v Martin* (1937)).

Martin could also sue Neil in negligence and under the Animals Act 1971. Under s 2(2) of the Act, it would have to be shown that:

(a) the damage is of a kind which the animal, unless restrained, was likely to cause or which, if caused by the animal, was likely to be severe;

(b) the likelihood of the damage or of its being severe was due to characteristics of the animal which are not normally found in animals of the same species, or are not normally found except at particular times or in particular circumstances; and

(c) those characteristics were known to that keeper.

As we are told that Neil owns the dog, then by s 6(3) he is the keeper.

The damage to plants will come under s 2(2)(a), and a tendency to eat plants is a characteristic not usually found in dogs and Neil is presumably aware of this characteristic, satisfying both s 2(2)(b) and 2(2)(c). Thus, all the requirements of s 2(2) have been met. There is no need to show any negligence on Neil's part (*Curtis v Betts* (1990); *Mirvahedy v Henley* (2002)).

Notes

Question 42

Oliver is employed as a salesman. He is calling on Peter's shop to sell them some office stationery, when he sees a gold watch on the floor. He picks it up and hands it to Peter, who takes his name and address. Some three months later, Oliver is passing Peter's shop, when he sees the watch in the window for sale. Oliver goes in and takes the watch from the window, but Peter grabs the watch from Oliver and there is a scuffle in which Oliver is injured.

Advise Oliver.

Would your advice differ if Oliver found the watch behind the counter of Peter's shop?

Answer plan

This is a relatively straightforward question of title to lost goods and also involves an element of trespass to the person.

The following points need to be discussed:

- Oliver's right to the watch as against Peter's;
- rules regarding supervening possession;
- necessary intention present in non-public part of shop; and
- the effect of Oliver being a trespasser.

Answer

Oliver will wish to sue Peter for conversion of the watch and for trespass to the person.

Conversion has been defined as being 'committed wherever one person performs a positive wrongful act of dealing with goods in a manner inconsistent with the rights of the owner' (*Maynegrain v Campafina Bank* (1984), *per* Lord Templeman). The tort is one of strict liability, in that, provided that the defendant intends to deal with the goods in a manner which is inconsistent with the owner's (or someone with a superior right to the goods) rights, the fact that the defendant is ignorant of these rights is no defence. So, for example, the innocent purchaser from a thief of stolen goods commits a conversion against the owner (*Moorgate Mercantile v Twitchings* (1977)). Hence, it follows that Peter has committed a conversion of the watch in offering it for sale.

To sue in conversion, Oliver must show that he had the right to possession (*Marquess of Bute v Barclays Bank* (1955)). The owner of the watch of course remains the owner but, as he has not claimed his property, the normal rule is that the finder, that is, Oliver in our case, has a right to it against everyone except the true owner, if he has reduced the goods into his possession. Thus, in *Armory v Delamirie* (1721), the finder of a jewel was held to be able to recover it from a jeweller to whom it had been handed and who refused to return it.

All this, however, assumes that the finder was the first person to reduce the goods into his possession and the possession counts as title (*The Winkfield* (1902)). However, we must decide whether someone other than Oliver had obtained earlier possession of the goods when Oliver found them, in which case, that person and not Oliver has the right to the goods. This can occur in two ways. First, if an employee finds goods in the course of his employment, the employee's possession is deemed to be that of his employer and the employee gains no possessory right against his employer (*Parker v British Airways Board*

(1982)). The important element here is that the goods must be found in the course of employment, that is, the employment must be the cause of the finding of the goods and not merely the occasion of the finding of the goods (*Byrne v Hoare* (1965)). We are told that Oliver is a salesman and that he calls on Peter's shop to sell some stationery. Oliver was undoubtedly going about his employer's business when he found the watch, but it seems that Oliver's employment was the occasion of his finding the watch, rather than the cause, and so it is submitted that Oliver's employer does not have a right of possession to the watch. Secondly, if goods are found on land not occupied by the finder, in certain situations, the occupier's occupation will confer upon him a possession of the lost goods, which is earlier in time than the finder and this previous possession can exist even though the occupier was unaware of the presence of the lost goods on his land. Such earlier possession will arise where the goods are buried on the land or attached to the land in such a manner as to suggest that the occupier is exerting exclusive control over the relevant area (*South Staffordshire Water Co v Sharman* (1896)). Where the goods are found just lying on the premises and the public have access to the premises, the finder generally has a superior right to the occupier (*Bridges v Hawkesworth* (1851); *Hannah v Peel* (1945)), unless the occupier has 'manifested an intention to exercise control over the building and things which may be in or on it' (*Parker v British Airways Board* (1982), *per* Donaldson MR). In *Bridges*, the finder of some cash in a shop was held to be entitled to it as against the shopowner, and the more modern case of *Parker* shows that the required intention is not easy to establish. In *Parker*, the finder of a bracelet in an airport lounge was held to be entitled to it as against the occupiers of the lounge. Thus, the weight of authority would allow Oliver a superior right to the watch as against Peter. By s 3 of the Torts (Interference with Goods) Act 1977, Oliver may obtain a court order to the watch or damages or delivery with the alternative of damages.

As Oliver is entitled to the watch, he is entitled to recover it from Peter using reasonable force if necessary to protect his property. As Peter has directly and intentionally applied hostile touching to Oliver's person, Oliver can sue Peter in battery (*Wilson v Pringle* (1987)) and in assault, if he was first put in reasonable fear of immediate physical contact.

If the watch was found behind the counter, Peter will find it easier to establish the intention described in *Parker*, as it would be easier to show that Peter intended to exercise control of the area behind the counter and any things in it. In addition, if Oliver was trespassing when he went behind the counter, as it was not part of the premises to which he was invited (*The Calgarth* (1927)), then as a trespasser he would acquire no rights as against the occupier (*Parker*).

It seems likely that Peter did evince the required intention in respect of the area behind the counter. Thus, he has a right to the goods as against Oliver and, when Oliver removed the watch from the window, Oliver was committing a conversion of the goods. Peter was entitled, therefore, to use reasonable force to protect his property, and he would be able to sue Oliver in battery and possibly assault as regards the ensuing struggle.

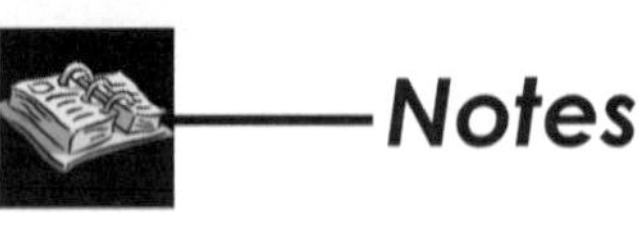

Notes

Question 43

George is walking along the road when he slips and falls. Frank sees this and goes to help George. While he is helping George to his feet, Ian comes along and, thinking that Frank is trying to rob George, Ian grips Frank by the arm and says, 'I am taking you to the police, you thief'. Frank struggles to free himself and pushes Ian to the ground. Ian sees a passing policeman and tells him that Frank has tried to rob George and has hit him (Ian). The policeman arrests Frank and takes him to the police station, where is he released after a few hours.

Advise Frank of the legal situation.

Answer plan

This question calls for a discussion of the various forms of the tort of trespass to the person, namely, assault, battery and false imprisonment, together with the defence of lawful arrest. The tort of defamation also needs to be considered.

The following points need to be discussed:

- Ian's liability to Frank for assault, battery and false imprisonment;
- possible defence of lawful arrest and the trap in *Walters v Smith* (1914);
- defamation by Ian on two occasions, and possible defence of qualified privilege;
- liability of policeman to Frank for assault, battery and false imprisonment and defence of lawful arrest;
- liability of policeman for defamation and defence of qualified privilege; and
- liability of Frank to Ian for assault or battery and defence of self-defence.

Answer

We need to advise Frank of any torts that may have been committed by Ian or the policeman, and of any liability that Frank may have incurred to Ian.

Considering Ian's behaviour first, Ian may have committed an assault upon Frank. An assault is an attempt or threat to apply force to a person whereby that person is put in fear of immediate physical contact. The test is an objective one, in that the claimant must have a reasonable belief that the defendant can carry out his purpose (*Thomas v NUM (South Wales Area)* (1985)). Ian may well have been guilty of an assault, but we are not given enough facts to be certain, for if Ian approached Frank from behind and gripped his arm before saying anything, then the tort of assault would not have taken place, as Frank would not have been put in fear of immediate physical contact.[1]

When Ian gripped Frank by the arm, he committed the tort of battery. Battery consists of a direct act of the defendant which causes contact with the claimant's body without the claimant's consent. The act must be both direct (*Scott v Shepherd* (1773)) and intentional (*Stanley v Powell* (1891); *Letang v Cooper* (1965)), and Frank would have no difficulty in showing either of the requirements. There must be some contact with the claimant's person, however trivial (*Cole v Turner* (1704)), and contact has clearly occurred. Finally, the touching must be hostile (*Wilson v Pringle* (1987)), which does not mean that Ian need show any ill will or malevolence, but he must show hostility. This requirement has been doubted by Lord Goff in an *obiter* statement in *F v West Berkshire Health Authority* (1989), but, as Lord Goff argued, a touching could amount to a battery in the absence of hostility and the fundamental question was whether the claimant consented to the touching. Even if Lord Goff's approach were to be used, Ian's act in gripping Frank's arm would constitute a battery.

We should next consider whether, in restraining Frank, Ian has committed the tort of false imprisonment, which consists of the total deprivation of the freedom of a person for any period of time, however short, without lawful justification (see *R v Bournewood Community and Mental Health NHS Trust* (1998) for a restatement by the House of Lords of the elements of false imprisonment). When Ian grips Frank by the arm, he deprives him of his freedom for a short period and so all the elements of this tort are present.

Given that, *prima facie*, these torts have taken place, we need to consider whether Ian has any defences available to him. The only defence relevant to Ian is available for all three torts, namely, the defence of lawful arrest under the provisions of the Police and Criminal Evidence Act (PACE) 1984. By s 24(5), where an arrestable offence has been committed, any person, that is, a private citizen or a police officer, may arrest without warrant anyone who is guilty of the offence or anyone who he has reasonable grounds for suspecting to be guilty of the offence. This means that where an offence has been committed, s 24(5) provides a defence where the person arresting arrests an innocent person. If an offence has not been committed, then, by s 24(6), a police officer may arrest without warrant anyone who he reasonably suspects to have committed an offence. Thus, the 1984 Act preserves the trap in *Walters v Smith* (1914), where it is necessary for a private citizen to prove that the offence in question has in fact been committed by someone. The case of *R v Self* (1992) is an example of the trap in action, where a person was reasonably suspected of a theft and was arrested by a private citizen. He resisted arrest and was later convicted of assault to resist lawful arrest, but acquitted of theft. It was held by the Court of Appeal that the conviction for assault to resist lawful arrest could not stand, because s 24(5) required, as a condition precedent, that an arrestable offence had been committed. Consequently, as the appellant had been acquitted of the alleged offence, a private citizen could not carry out an arrest under s 24(5), which did not operate under these circumstances. As Frank has not committed the arrestable offence of which Ian suspected him, his arrest of Frank is not lawful and Ian cannot avail himself of the defence of lawful arrest. The fact that Ian might have had reasonable grounds to suspect Frank of the offence is irrelevant (s 24(5); *R v Self*) if the offence has not in fact been committed.

We should also advise Frank whether, in struggling with Ian and pushing him to the ground, Frank himself has committed any torts against Ian. From our discussion above, it is clear that Frank may be guilty of assault and/or battery, although on the facts that we are given there does not appear to have been any false imprisonment of Ian. The defence that would be appropriate for Frank to invoke is that of self-defence, for, as Ian's attempted arrest is not lawful, Frank is entitled to protect himself against Ian's assault and battery. The rule is that, in self-defence, the steps taken to protect oneself must not be out of all proportion to the harm threatened (*Lane v Holloway* (1968)) and, provided that Frank satisfies this criterion, he will have a valid defence to any action by Ian.

In calling Frank a thief, Ian has defamed Frank. To succeed in an action for defamation, Frank will have to show that Ian made a defamatory statement that could reasonably be understood to refer to him, and that this statement was published to a third party. As the statement in question was spoken, it was in the form of slander and normally, in slander, the claimant will have to prove special damage. However, where the statement imputes a crime punishable by imprisonment, the statement is actionable *per se* (*Hellwig v Mitchell* (1910)). The test for the defamatory nature of a statement is whether it would lower the claimant in the estimation of right-thinking members of society (*Sim v Stretch* (1936)), which Ian's allegation does. It also seems clear, from what we are told, that Ian could not avail himself of the defence that the slander was mere abuse (*Parkins v Scott* (1862); *Lane v Holloway* (1968)), as it could be intended to be taken seriously. The statement obviously refers to Frank and has been published to a third party, namely, George. Thus, all the elements of defamation are present, and there are no defences on which Ian could rely in respect of this statement. When Ian makes his allegations to the policeman, the elements of defamation are again present, but Ian will be able to rely on the defence of qualified privilege. This defence operates where, *inter alia*, one person has a legal, social or moral duty to make a statement to another and that other has a corresponding duty to receive it (*Watt v Longsden* (1930)). Ian has such a duty as regards criminal acts and the policeman has a duty to receive such statements.

Thus, Ian could rely on the defence of qualified privilege, which Frank could only destroy by showing that Ian was actuated by malice when he made the statement, which in the circumstances seems unlikely.

Recently, the Court of Appeal in *Kearns v General Council of the Bar* (2002) held that the duty-interest test was not always useful, and that it would be more helpful to distinguish between cases where the communicating parties were in an existing and established relationship and cases where no such relationship had been established and the communication was between strangers. The court held that privilege would attach much more readily in the established relationship situations. It is submitted, however, that *Kearns* was not intended to do away with qualified privilege in the well-established situation of a member of the public giving information regarding an alleged offence to a police officer.

When the policeman arrests Frank, he may have committed an assault or battery, but he can rely on s 24(6) of PACE 1984 as a defence. Section 24(6) allows a police officer to make a lawful arrest where he reasonably suspects that someone has committed an arrestable offence, and the policeman has such reasonable grounds. This lawful arrest will also, of course, be a valid defence to any action for false imprisonment. Likewise, if the policeman makes any statements to his colleagues at the police station regarding Frank's alleged criminal acts, such statements would be protected by the doctrine of qualified privilege as both the policeman and his colleagues have the required interests, as discussed above.

Thus, Frank could sue Ian for assault, battery, false imprisonment and defamation, but could not sue the policeman. Frank could not be sued by Ian in respect of the struggle. Frank could not sue Ian for false imprisonment while Frank was in police custody, as Ian merely gave information to the police who effected Ian's arrest and detention (*Davidson v Chief Constable of North Wales* (1994)).

Think point

1 If Ian said 'I am taking you to the police, you thief' before gripping Frank, it is submitted that this would constitute an assault. In *R v Meade and Belt* (1823), it was said that words cannot amount to an assault, but this case has been subject to both academic and judicial criticism (*R v Wilson* (1955)), although it is supported by Buckley and Heuston (eds), *Salmond and Heuston on the Law of Torts*, 21st edn, 1996.

Notes

13 Economic Torts

Introduction

Questions on the economic torts are popular with examiners, as not only is it an important topic, but it has also been the subject of important developments. In addition, the exact scope of some of the torts is subject to some uncertainty, calling for a careful analysis of some decisions of the courts. This is a complex area, and candidates should only attempt questions if they have a reasonably good and up to date knowledge of the topics.

Checklist

Students must be familiar with the following areas:

(a) conspiracy involving lawful and unlawful acts, defences to lawful act conspiracy;

(b) inducing a breach of contract – mental state required and defences;

(c) intimidation;

(d) interference with trade by unlawful means – note the uncertainty regarding the ingredients and extent of this tort; and

(e) deceit, malicious falsehood and passing off.

Question 44

Alfred runs a grocery business, which supplies packed lunches to several nearby factories. Brian and Charles run a nearby sandwich bar and feel that Alfred's competition, as he is able to purchase foodstuffs at reduced prices via his grocery business, is unfair. They therefore falsely tell the personnel officer at the factory that they believe that the local food inspector is unhappy with the state of hygiene at Alfred's premises, and as a result the factory ceases to use Alfred as a supplier. When Alfred becomes aware of the activities of Brian and Charles, he immediately informs David, who supplies bread to both Alfred and Brian and Charles, that he (Alfred) feels he cannot do business with a person who supplies his rivals and that David must choose to do business with Alfred or Brian and Charles, but not both.

Advise Alfred as to the legal situation.

Answer plan

This is a typical problem on the economic torts, in that the facts disclose a range of possible causes of action, which must all be discussed. In this type of question, the student should not neglect the possibility of an action arising in the traditional tort areas of deceit, malicious falsehood and passing off.

The following points need to be discussed:

- Alfred's competition via his grocery business;
- Brian and Charles' report to the factory;
- conspiracy by unlawful means;
- inducement to breach of contract;
- interference with trade by unlawful means;
- malicious falsehood;
- Alfred's dealing with David;
- inducing breach of contract; and
- intimidation.

Answer

Alfred should be advised as to whether Brian and Charles have committed any torts, and whether he himself has committed any torts in his dealings with Brian and Charles and with David.

As regards Alfred's competition with Brian and Charles via his grocery business, this does not give rise to any course of action. Competition, however vigorous or unfair, is not unlawful (*Mogul Steamship v McGregor Gow* (1892)) and, despite the modern development in the area of the economic torts, Hoffmann J held that there is no tort of unfair trading (*Associated Newspapers Group v Insert Media* (1988)).

When, however, Brian and Charles made their statement to the personnel officer of the factory, a number of possible courses of action arise.

First, Brian and Charles may be guilty of conspiracy to commit an unlawful act. Conspiracy has been defined as the agreement of two or more persons to do an unlawful act or a lawful act by unlawful means (*Mulcahy v R* (1868), *per* Willes J). Although the word 'agreement' was used in this definition, the word 'combination' is now preferred, since there is no need for a contractual agreement, merely that the persons conspire together with a common purpose (*Belmont Finance Corp v Williams Furniture* (1980)). The conspiracy must cause damage to the claimant, for the tort is not actionable *per se*, and this has clearly happened, as the factory has ceased to trade with Alfred. Before we can conclude that all the ingredients of the tort are present, we should consider what constitutes an unlawful act. Although not all wrongful or illegal acts will support an action for unlawful means conspiracy – see *Michaels v Taylor Woodrow Developments Ltd* (2001) – a tort is unlawful for this purpose (*Sorrel v Smith* (1925)) and here Brian and Charles are committing the tort of malicious falsehood and also conspiring to induce a breach of contract. In the House of Lords' decision in *Lonrho v Al-Fayed* (1991), it was held that the tort of conspiracy to injure could be established by showing that an intent to injure the claimant's business interest was the predominant purpose, even though the means were lawful and would not have been actionable, if carried out by an individual, or by showing that unlawful means were used. However, where there was an intent to injure, and unlawful means were used, it was no defence for the defendants to show that their predominant purpose was to protect their own interest rather than to injure the claimant's business – it was sufficient that they had used unlawful means to constitute the tort. Hence, Brian and Charles could not avail themselves of the defence that their predominant purpose was to defend their interest, rather than to injure Alfred's interests.

Secondly, Brian and Charles may have committed the tort of inducing a breach of contract (*Lumley v Gye* (1853)). Following the classification of Jenkins LJ in *DC Thomson v Deakin* (1952), Brian and Charles' action is that of direct persuasion to breach the contract between the factory and Alfred. The court needs to find some persuasion to breach the contract and, in *Square Grip Reinforcement v MacDonald* (1968), Lord Milligan stated that where a defendant was 'desperately anxious' to achieve a particular result, the court would be likely to construe a statement made by the defendant as persuasion, and that appears to be the case here. It must also be shown by Alfred that Brian and Charles knew of the contract between Alfred and the factory and acted with the intention of bringing about a breaking of this contract. On the facts of the problem, Alfred would seem to have little difficulty here and should be able to establish that this tort has been committed.

Thirdly, Brian and Charles may have committed the tort of interference with trade by unlawful means. This now seems to be established as a tort in its own right (*Merkur Island Shipping v Laughton* (1983); *Hadmor Productions v Hamilton* (1982)). In *Lonrho v Al-Fayed* (1991), the Court of Appeal held that it was not an essential ingredient of the tort that the defendant's predominant purpose was to injure the claimant, rather than to further his own interest, but it was necessary to prove that the action was directed against the claimant or intended to harm the claimant. Alfred should have no problem in establishing this, but there seems to be considerable uncertainty as to just what constitutes unlawful means. In *Lonrho v Shell Petroleum* (1982), the breach of a penal statute was held to be insufficient to found a cause of action and, in *Chapman v Honig* (1963), a criminal contempt of court was also held to be insufficient. However, in *Acrow v Rex Chainbelt* (1971), such a contempt was held to be sufficient and, in *Associated British Ports v TGWU* (1989), the Court of Appeal held that a non-actionable breach of a statute could constitute unlawful means if it was coupled with an intent to injure the claimant. In the light of these cases, it is submitted that Brian and Charles have committed the tort of unlawful interference with trade, allowing Alfred to sue.

Fourthly, Brian and Charles have committed the tort of malicious falsehood. To establish this, Alfred must prove that Brian and Charles have made a false statement to someone other than Alfred; that that statement was made maliciously; and that the statement has caused damage to Alfred. There is no problem in showing that the statement is a false statement of fact – it is clearly not opinion or a trade puff (see the judgment of Walton J in *De Beers Products v Electric Co of New York* (1975)). To show malice, Alfred must prove that Brian and Charles acted out of spite or with a desire to injure him and, on the facts we are given, Alfred should have no problem here. Finally, Alfred must show that he has suffered damage which can be shown by the loss of business (*Ratcliffe v Evans* (1892)). In addition, by s 3(1)(b) of the Defamation Act 1952, if the words are calculated to cause pecuniary damage to the claimant in any trade or business carried on by him, there is no need to prove special damage.

Finally, Brian and Charles may also have defamed Alfred, as the intimation that he runs a food business which is unhygienic is defamatory. The statement refers to Alfred and was published to a third party. Although it was an oral statement and is thus slander, it is calculated to disparage Alfred in his business and thus no special damage need be proved by Alfred (s 2 of the Defamation Act 1952).

Turning now to Alfred's dealings with David, Alfred himself has committed a number of torts. First, Alfred has committed the direct form of the tort of inducing a breach of contract, as all the elements identified in the earlier discussion are present. Secondly, Alfred may have committed the tort of interference with trade by unlawful means. Again, the problem of what constitutes unlawful means arises, and whether a breach of contract is sufficient when combined with an intent to harm Brian and Charles' business. Finally, Alfred may have committed the tort of intimidation, that is, a threat to a third party that the defendant will use some unlawful means against the third party unless the third party does or refrains from doing some act he is entitled to do, and consequently the claimant suffers loss (*Rookes v Barnard* (1964)). Alfred has issued a threat to David, but the question is whether he has used unlawful means. In *Rookes*, it was held that, for the purpose of this tort, a breach of contract constitutes unlawful means. Provided that David submits to Alfred's threat and Brian and Charles suffer damage as a result, all the ingredients of intimidation are present.

Notes

Question 45

George owns a small engineering business, and replies to a tender issued by Gamma Manufacturing plc for the supply of some components. He mentions this to Henry, who owns a similar business and, when Henry says that he might also tender for the components, George tells Henry that Gamma have a reputation for delay in paying their suppliers. Gamma does not have such a reputation, and George knows that Henry's cashflow situation is delicate and that he cannot afford to deal with very slow payers. As a result, Henry does not submit a bid to Gamma and George obtains the contract.

Advise Henry.

What would be the legal situation if, prior to Gamma contracting with George, Henry called on Gamma and claimed that his components were superior to those that George supplied and thus obtained the contract, when in fact Henry knows that his components and George's are of a similar quality?

Answer plan

This is a question on economic torts which is a little unusual, in that it gives greater emphasis to the 'traditional' economic torts of deceit and malicious falsehood than to the more 'modern' torts of conspiracy, inducing a breach of contract, intimidation and interference with trade by unlawful means. These 'traditional' economic torts are still important and, when candidates encounter a question that seems to be testing the more 'modern' economic torts, the facts of the problem should be read carefully to see if they disclose the possible existence of the 'traditional' economic tort.

The following points need to be discussed:

- elements of conspiracy, inducing a breach of contract and intimidation not present;
- interference with trade by unlawful means by George;
- deceit by George;
- malicious falsehood by Henry;
- discussion of fact versus puff;
- inducement; and
- damage and s 3 of the Defamation Act 1952.

Answer

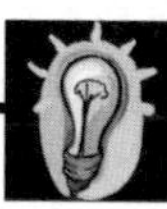

We need to advise Henry as to whether he has any course of action in respect of George's false statement concerning Gamma. Henry can have no cause of action in conspiracy, as George has acted alone. If he had conspired with his company (assuming that his business has been incorporated as a company), that would be sufficient (*Belmont Finance Corp v Williams Furniture* (1980)), but there is no evidence to that effect. George cannot be liable for inducing a breach of contract for, when George made the statement to Henry, there was no contract between Henry and Gamma. Neither can George be liable in the tort of intimidation, as that requires a threat to be made to a third party (*Rookes v Barnard* (1964)), and the statement, which does not seem to be a threat anyway, was made to the claimant, Henry.

However, George may have committed the tort of interference with trade by unlawful means. This now seems to be established as a tort in its own right (*Merkur Island Shipping v Laughton* (1983); *Hadmor Productions v Hamilton* (1982)). In *Lonrho v Al-Fayed* (1991), the Court of Appeal held that it was not an essential ingredient of the tort that the defendant's predominant purpose was to injure the claimant, rather than to further his own interests, but it was necessary to prove that the act was directed against the claimant or intended to harm the claimant. Henry should have no problem in establishing this, but there seems to be a considerable uncertainty as to just what constitutes unlawful means. In *Lonrho v Shell Petroleum* (1982), the breach of a penal statute was held to be insufficient to found a cause of action and, in *Chapman v Honig* (1963), a criminal contempt of court was also held insufficient. However, in the later case of *Acrow v Rex Chainbelt* (1971), such a concept was held sufficient and, in *Associated British Ports v TGWU* (1989), the Court of Appeal held that a non-actionable breach of a statute could constitute unlawful means if it was coupled with an intent to injure the claimant. In the light of these later cases, it is submitted that George has committed at least the tort of unlawful interference with trade, allowing Henry to sue.

Finally, George has committed the tort of deceit. This tort has five elements (*Pasley v Freeman* (1789)). First, the defendant must have made a false representation of fact. Henry's statement that Gamma are slow payers is a representation and we are told that it is false. The statement seems to be a representation of fact. Henry could argue that, as he has stated that Gamma have a reputation as slow payers, this is not a fact, but an opinion, as in *Bisset v Wilkinson* (1927). However, if Henry could not have honestly held that opinion or was warranting that he knew facts to justify this opinion, his statement will be treated as one of fact (*Smith v Land and House Property Corp* (1884)). We shall treat George's statement as being a statement of fact and we are told that it is false. Secondly, the defendant must know that the statement is false, has no belief in its truth or be recklessly careless whether it be true or false (*Derry v Peek* (1889)). In *Angus v Clifford* (1891), Bowen LJ stated in the Court of Appeal that 'careless' meant indifference to the truth or wilful disregard of the importance of the truth.

In view of George's knowledge of Henry's cashflow situation, it seems likely that at the very least George was recklessly careless whether his statement was true or false.

Thirdly, the defendant must have intended that his statement be acted upon and, as George made the statement directly to Henry, no problem arises here. Fourthly, the defendant's false statement must have been relied upon by the claimant. It need not be the sole or decisive factor in causing the claimant to act as he did (*Edgington v Fitzmaurice* (1885)); it is sufficient if it was one of the reasons. In *McCullagh v Lane Fox* (1994), it was held by the High Court that in the tort of deceit it was enough that the claimant's judgment was influenced by the statement. Although *McCullagh* was upheld by the Court of Appeal on different grounds, this point was not disturbed. Provided that Henry did not undertake his own investigations into Gamma's speed of paying, and relied on those investigations (*Atwood v Small* (1838)), it is irrelevant that Henry had the means to discover that these statements were false (*Redgrave v Hurd* (1881)). Finally, Henry must show that he suffered damages as a result of George's false statement, which he has done by losing the chance of the contract.

In the event of Henry claiming to Gamma that his components were superior to George's, when Henry knew that this was not true, Henry commits the tort of malicious falsehood.

Malicious falsehood consists of the defendant making a false statement, with malice, to a third person, as a result of which the claimant suffers damage. If George wishes to sue Henry in this tort, he must first show that Henry made a false statement to a third party. The false statement must be one of fact and not mere opinion or puff. Where a defendant makes a statement which boosts his own goods, that is a mere puff (*White v Mellin* (1895)). However, where the defendant makes disparaging remarks about the claimant's goods, the statement is more likely to be treated as a statement of fact (*Lyne v Nichols* (1906); *De Beers Products v Electric Co of New York* (1975)). This is particularly so where the defendant's statement is intended to be taken seriously (*Lyne*; *De Beers*) because, for example, it quoted facts or alleged facts (*De Beers*). In *De Beers*, Walton J held that the test to apply was whether a reasonable person would take the defendant's statement as a serious statement. Thus, in the present case, if Henry merely said, 'my components are better than George's', that would be a mere puff; if Henry said, 'my components are better than George's because ...', this statement would probably not be treated by the court as a mere puff.

George must also show that Henry's statement was made maliciously but, where the defendant makes the statement knowing it is false, he is acting maliciously (*Greers v Pearman and Corder* (1922)). As we are told that Henry knows that his statement is untrue, there is no problem regarding malice. Finally, Henry must show that he has suffered damage as a result of George's statement. As he has lost the contract with Gamma, this is enough but, by s 3(1)(b) of the Defamation Act 1952, where the words are calculated to cause pecuniary damage to the claimant in so far as any trade or business carried on by him, he need not prove special damage.

Thus, depending on the exact words used by Henry, George will be able to sue Henry in malicious falsehood.

Notes

Question 46

'If a person intentionally interferes with another's business, that will constitute a tort, but not if such intent is missing.'

Discuss whether the above statement is an accurate summary of the current legal position.

Answer plan

This question calls for a discussion of the role of intent in what are commonly called the economic torts. In answering this question, students should not only cover those torts generally grouped in the textbooks under the heading of the economic torts, but also the torts of deceit, malicious falsehood and passing off.

The following points need to be discussed:

- conspiracy by unlawful acts and lawful means;
- inducing breach of contract;
- intimidation;
- interference with trade by unlawful means;
- deceit;
- malicious falsehood; and
- passing off.

Answer

Although we shall attempt to show that the statement under discussion does represent accurately the legal position, it should not be thought that any act which is done with the intent of damaging the business of another is automatically a tort. Such a proposition was expressly rejected by Hoffmann J in *Associated Newspapers Group v Insert Media Ltd* (1988), that is, English law does not recognise a tort of unfair trading.

We shall first examine the statement as it might apply to the tort of conspiracy. A conspiracy has been defined as consisting 'not merely in the intention of two or more, but in the agreement of two or more to do an unlawful act, or to do a lawful act by unlawful means' (*Mulcahy v R* (1868), *per* Wiles J). Taking the first type of conspiracy, it can be seen from the definition that an unlawful act is required. So, in *Mogul Steamship v McGregor Gow* (1892), where the defendants attempted to obtain a monopoly in the tea trade by reducing their prices to drive the plaintiffs out of business, it was held that the plaintiffs could not sue the defendants in conspiracy, as the defendants had committed no unlawful act against the plaintiffs. Similarly, in *Allen v Flood* (1898), the plaintiffs failed as they could not show an unlawful act on the part of the defendants. The fact that the defendants in *Allen* had acted out of spite or malice was irrelevant, because such considerations could not turn a lawful act into an unlawful act. The question that obviously arises now is what constitutes an unlawful act. In *Michaels v Taylor Woodrow Developments Ltd* (2001), it was held that not all wrongful or illegal acts will suffice for an unlawful means conspiracy, but that a crime, a tort and probably a breach of contract will suffice. The House of Lords has considered the necessary intent for this tort and, in *Lonrho v Al-Fayed* (1991), held that where unlawful acts or means are used, it is not necessary to show that the defendants had, as their predominant purpose, an intent to injure the claimant – an intent to injure coupled with unlawful means is sufficient (*per* Lord Bridge). Let us turn next to the second type of conspiracy, the lawful act conspiracy. This anomalous tort has its origins in the decision of the House of Lords in *Quinn v Leatham* (1901), where it was held that where persons conspired to inflict unjustified harm on another, a cause of action arose. This tort is recognised as anomalous, for it means that two or more persons acting together can turn an act which is lawful into one which is unlawful. Also, in *Lonrho v Shell Petroleum* (1982), the House of Lords held that it was anomalous and was not to be extended. The necessary intent for this tort was considered by the House of Lords in *Crofter Hand Woven Harris Tweed v Veitch* (1942), where it was held that no action would lie unless the predominant purpose for the defendants' actions was to injure the claimants. Thus, in the case of *Yukong Line Ltd of Korea v Rendsburg Investments Corp of Liberia* (1998), the claimant failed, as the predominant purpose of the defendant was to advance his own interest rather than injure the claimant. As in *Crofter*, the main purpose of the defendants' actions was to protect the interests of their members, and the claimants failed. In this tort, the pursuit of self-interest is a defence (*Crofter*) and it has been held that a justified purpose may also be a valid defence (*Scala Ballroom v Ratcliffe* (1958)). Thus, it can be seen that for this anomalous form of the tort of conspiracy, intent to injure is an essential ingredient and that it needs to be the predominant intent. If the intent to injure is present, but it is not the main intent, that is insufficient intent to constitute the tort.

Another way in which a person may interfere with another's business is through the tort of inducing a breach of contract. This tort has its origins in *Lumley v Gye* (1853) and, in *Thomson v Deakin* (1952), it was stated that the tort could take one of three forms: namely, direct persuasion of a contracting party to break his contract; where the defendant prevents performance of the contract by some direct and wrongful means; and, finally, where A induces a third party to break his contract with B, so that B is unable to perform his contract with the claimant. In practice, it can be difficult to distinguish between these various forms (*Stratford v Lindley* (1965)). In any event, it must be shown that the defendant had the required intent – it is necessary to show that the defendant had both knowledge of the contract and the intent to procure a breach of that contract. The courts seem ready to infer that the defendant knew that his actions would lead to a breach of contract (*Merkur Island Shipping v Laughton* (1983)), and the intent to procure a breach of contract can be shown by proving that the defendant was reckless as to whether a contract was breached or not (*Emerald Construction v Lowthian* (1966)). Thus, intent is a necessary ingredient of this tort, and its presence seems reasonably easy to demonstrate in appropriate circumstances. However, even where this intent is present, it does not mean that liability will automatically arise. In particular, where direct pressure is brought to bear on a stranger to the contract and the effect on the contracting party is indirect, it is necessary to show that some unlawful means had been used against the stranger to the contract if the conduct is to be actionable (*Middlebrook Mushrooms Ltd v Transport and General Workers Union* (1993)). As with lawful means of conspiracy, various defences are available to the defendant. These defences were discussed by the Court of Appeal in *Edwin Hill v First National Finance Corp* (1989) and are, first, where the contract interfered with is inconsistent with a previous contract with the interferer (*Smithies v National Association of Operative Plasterers* (1909)) and, secondly, where there is a moral duty to intervene (*Brimelow v Casson* (1924)). Thus, again, intent to damage another is a necessary, but not necessarily sufficient, condition for liability.

Another route by which a person may interfere with the business of another is through the tort of intimidation. This tort was analysed by the House of Lords in *Rookes v Barnard* (1964). The ingredients are the threat of some unlawful act to a third party, to which threat the third party submits, and this action by the third party causes damage to the claimant. It seems difficult to imagine this tort taking place without intent on the part of the person making the threat, so again intent is required. In the present state of the law, it is an undecided point whether a defence of justification exists to this tort. Indeed, while it would not seem possible to justify an unlawful act, which is an essential element of this tort, Lord Denning has suggested that this defence could be available in appropriate circumstances (*Morgan v Fry* (1968); *Cory Lighterage v TGWU* (1973)), so again it may be that intent is a necessary, but not necessarily sufficient, element to establish liability.

Interference with another's business may also occur via the recently recognised tort of interference with trade by unlawful means. This activity was recognised as a tort in its own right by the House of Lords in *Merkur Island Shipping v Laughton* (1983). In *Lonrho v Al-Fayed* (1991), the Court of Appeal held that it was not necessary to prove that the predominant purpose of the defendant was to injure the claimant, but that it was enough to show that the unlawful act of the defendant was directed against the claimant or was intended to harm him. This tort is still at an early stage of its development by the courts, but it seems that an intent to injure the claimant is an essential element, although the scope of the defences to this tort are still rather obscure.

It should be noted that in *Cruickshank v Chief Constable of Kent County Constabulary* (2002), the Court of Appeal held that although an action for interference with contractual relations could be brought against a public official, in all but extreme cases the tort of misfeasance in public office would afford all appropriate remedies.

In addition to the standard economic torts that we have considered above, a person may interfere with the business of another in several ways which can give rise to certain well-established torts, which we shall now consider. Thus, a person may wilfully or recklessly make a false statement to another with the intent that the other shall act in reliance on it and, if that other relies on it and suffers damage thereby, the person making the statement is liable in the tort of deceit. A necessary ingredient of the tort of deceit is an intent on the part of the defendant that the statement be acted upon and, without such an intent, the tort is not constituted. To this extent, the statement under discussion accurately reflects the law.

A person may also make a false statement to someone other than the claimant, as a result of which the claimant suffers damage. This constitutes the tort of malicious falsehood and, in *Ratcliffe v Evans* (1892), it was said that liability would arise where the false statement was *inter alia* 'calculated to produce ... actual damage'. Thus, it can be seen that an essential element of this tort is an intent to injure the claimant.

Finally, a person may pass off his goods as being those of someone else. In *Erven Warnink v Townend* (1979), Lord Diplock stated that an essential element of this tort was an intent to injure the business or goodwill of another.

Overall, therefore, it can be seen that the statement under discussion is an accurate summary of the current legal position, in that the intent in question is always required.

Notes

14 Remedies

Introduction

The most important remedy in tort is damages, and questions involving damages for personal injury or death are often set by examiners. Such questions may take the form of a general essay or a problem question, in which details are given of the claimant's salary and family responsibilities. In the latter type of question, candidates are not expected to produce detailed calculations of damages, but rather to indicate and discuss the particular heads of damage which are recoverable and how they would be calculated.

Checklist

Students must be familiar with the following areas:

(a) types of damages – nominal, contemptuous, general, special damages, special damage (that is, actual loss which must be proved if the tort is not actionable *per se*), aggravated and exemplary;

(b) damages for personal injury – pecuniary and non-pecuniary loss;

(c) damages for death; and

(d) ss 2, 11 and 33 of the Limitation Act 1980.

Question 47

James is crossing the road when he is injured due to the negligent driving of Ken. As a result of this accident, James, who is married with two young children, will be confined to a wheelchair for the rest of his life. Explain how a court would assess what damages James should receive from Ken.

If James were to die one year after the accident, and before the trial, how would the damages then be assessed?

Answer plan

Although this is written in the form of a problem, it is in fact a directed essay on the calculation of damages for personal injury and death. It requires a consideration of the various heads of damage under which James could recover, but not actual estimates of the amount recoverable.

The following points need to be discussed:

- object of damages;
- damages for the claimant – the various heads of pecuniary loss and deductions;
- the various heads of non-pecuniary loss; and
- damages for death.

Answer

The object of awarding damages in tort is to put the claimant as far as money can do so in the position as if the tort had not happened. Thus, as a general rule, if, as a result of the accident, James has lost money or will have to spend money he otherwise would not have had to spend, he can recover in respect of these sums.

If we apply this general principle to the pecuniary loss that James has suffered, we can see that the first thing James has lost is wages, as he is now confined to a wheelchair (assuming for the present that James ceases to be paid any salary by his employer from the date of the accident). James will have lost a certain amount of wages up to the date of trial and this is calculated using his net wages as a basis, as James has only lost his take-home pay, not his gross pay. For further loss, the problem is more difficult, due to uncertainties of future income, life expectancy, etc. The court will calculate James' net annual loss and multiply that by a figure based on the number of years the loss is likely to last, the multiplier. The multiplier is not simply the duration of the disability, but a lower figure with a maximum value of around 25 (see *Wells v Wells* (1998)) to take account of the fact that James has received the money as a lump sum, rather than over a period of years. The multiplier is also designed to take account of the 'general vicissitudes of life'. It should also be noted that any award will be final and, should James' condition worsen, he will not normally be able to go back to court to claim any added sums (*Fitter v Veal* (1701)). Thus, it is essential to wait until James' medical condition stabilises before any trial. Section 32A of the Supreme Court Act 1981 does allow a provisional award to be made with the right to additional compensation should the condition worsen, but s 32A has been given a somewhat restrictive interpretation by the High Court in *Willson v Ministry of Defence* (1991).

An obvious problem for James is the effect of future inflation. No especial protection is given in respect of this (*Lim Poh Choo v Camden Health Authority* (1980)), but recently courts have begun to approve 'structured settlements', in which part of the sum payable to the claimant is invested by the defendants in an annuity which can provide an index-linked annual sum for the rest of the claimant's life (*Kelley v Dawes* (1990); s 2(1) of the Damages Act 1996).

It may be that, as a result of the accident, James has a reduced expectation of life. If so, James can recover the earnings he would have received during the years he has lost due to his reduced expectation of life (*Pickett v British Rail Engineering* (1980)) although, following the general principle of damages in tort, James' living expenses must be deducted (*Harris v Empress Motors* (1983)).

James will also be compensated for any loss of pension rights that accompanies his loss of salary.

James can claim in respect of any future expenses he will be put to as a result of the accident. Thus, James can recover for nursing care, and this may be obtained privately even if it is available under the NHS (s 2(4) of the Law Reform (Personal Injuries) Act 1948). James can also claim for any changes necessary to his accommodation, for example, the provision of wheelchair ramps, additional costs of lighting or heating and future costs of a gardener, tradesmen, etc, if James did these jobs himself and now cannot do so (see, for example, *Willson*).

It may well be the case that James receives compensation from a person other than the tortfeasor, and deduction from the previous amounts may have to be made to prevent double recovery.

The Social Security (Recovery of Benefits) Act 1997 provides that no deduction for social security benefits is to be made against awards for pain and suffering, and that specified benefits only may be deducted from awards for loss of earnings, cost of care and loss of mobility.

For other benefits, the general rule is that a benefit received by the claimant is only deducted where it truly reduces the loss suffered (*Parry v Cleaver* (1970)). Hence, sick pay or wages paid during the period following the accident are deducted, but not any insurance sums that James receives (*Bradburn v Great Western Railway* (1874)) or charitable donations, ill health awards or higher pension benefits (*Smoker v London Fire and Civil Defence Authority* (1991)).

James will of course also suffer non-pecuniary loss. First, there will be the pain and suffering that James has endured and will suffer in the future. Also, if as a result of the accident James has suffered a loss of expectation of life and is aware of this, then, by s 1(1)(b) of the Administration of Justice Act 1982, the court is required to take this into account when assessing damages. Next, James will be compensated for any loss of amenity, that is, his capacity to engage in pre-accident activities, and this award may be made even if he is in a coma (*West v Shepherd* (1964); *Lim Poh Choo*). James will also be compensated for the injury itself and, to obtain some consistency in this respect, a listing of awards is made in Kemp and Kemp, *The Quantum of Damages*. James should be advised that the Court of Appeal recently held in *Heil v Rankin* (2000) that awards for non-pecuniary loss were too low and that for the most severe injuries, the awards should be increased by about one-third.

Finally, James will be awarded interest on his damages in respect of losses up to the date of trial under s 35A of the Supreme Court Act 1981. For pecuniary loss, the interest rate is one-half the short term interest rate from the date of accident to the date of trial: *Jefford v Gee* (1970); *Cookson v Knowles* (1979). For non-pecuniary loss, the rate is 2% from the date of service of the writ to the date of trial: *Wright v British Railways Board* (1983).

If James dies before trial as a result of the accident, two causes of action arise. First, under s 1 of the Law Reform (Miscellaneous Provisions) Act 1934, all causes of action vesting in the deceased survive for the benefit of his estate. The damages which the estate can claim are assessed in a similar way to those in a personal injuries claim, except that a claim for lost earnings in the lost years can only be brought by a living claimant (s 4(2) of the Administration of Justice Act 1982). Secondly, an action may be brought under the Fatal Accidents Act 1976 by James' dependants. The dependants can claim a fixed sum of £7,500 for bereavement by a spouse for loss of a spouse or by parents for loss of a child, funeral expenses (if not paid by the estate) and actual and future pecuniary loss. This is calculated by assessing the dependency of the deceased, which is normally found by taking the deceased's net earnings, deducting a sum for his personal and living expenses and multiplying this sum by the duration of the dependency (which is calculated on a similar basis to the multiplier in personal injury cases). In assessing dependency and duration, any chance of a widow remarrying is to be ignored (s 3(3) of the Fatal Accidents Act 1976) (but not the chance that the parties might have divorced (*Martin v Owen* (1992))). Also, by s 4 of the 1976 Act, any benefits accruing as the result of the death are to be disregarded. Thus, for example, any widow's pension paid by James' employers to his widow is to be disregarded (*Pidduck v Eastern Scottish Omnibus* (1990)).

Notes

Question 48

It is a general rule of law that damages are awarded to compensate the claimant, rather than to punish the defendant. Are there any situations where a claimant could make a profit out of the damages awarded to him?

Answer plan

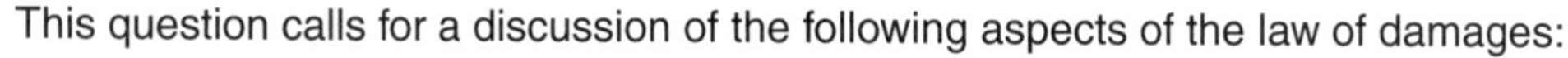

This question calls for a discussion of the following aspects of the law of damages:

- aggravated damages;
- exemplary damages;
- non-deduction of insurance sums;
- possible double compensation under the Fatal Accidents Act 1976; and
- damages in defamation.

Answer

The general principle governing an award of damages in tort is to put the claimant in the position he would have been in had the tort not occurred, as far as this can be done by an award of money. In some cases, for example, damages for negligent misrepresentation, the loss may be purely financial and it may be possible to calculate this loss precisely. However, in many situations, this will not be possible. For example, in personal injury cases, the loss of wages suffered by the claimant can be calculated exactly, but such a calculation is impossible as regards a broken thigh that caused an absence from work. Similarly, precise calculations of damages in (say) nuisance or trespass will generally not be possible. A claimant must expressly plead any special damage that he has suffered, for example, medical expenses or loss of wages, and will in addition be awarded general damages which are not quantified in the statement of claim, but are assessed by the court. These general damages attempt to compensate the claimant for the non-financial consequences that have flowed from the tort, and represent an estimate by the court, in money terms, of the claimant's loss. By definition, therefore, these general plus special damages, together with any interest awarded, should match the loss suffered by the claimant.

However, in certain circumstances, the court may make an award of aggravated damages. These damages are still regarded as compensatory damages, in that they are awarded to compensate the claimant for loss that he has suffered, rather than to punish the defendant. Aggravated damages may be awarded where the defendant's conduct caused injury to the feelings or pride of the claimant. In *Archer v Brown* (1985), it was stated that sums awarded in respect of aggravated damages should be moderate (see also *W v Meah* (1986)). An example of circumstances that might justify an award of aggravated damages can be seen in *Marks v Chief Constable of Greater Manchester* (1992). In *Marks,* the Court of Appeal held that a Chief Constable's conduct in persisting in a denial of liability in a civil action, despite comments which had been made by a recorder in criminal proceedings against the claimant as to conflicting police evidence, was capable of aggravating the claimant's damages should she be successful in her civil case, and might be grounds for an award of additional damages. It should be noted that in *Kralj v McGrath* (1986), it was held that medical negligence cases were not appropriate for an award of aggravated damages, but that, rather, the general damages should be increased to take into account the fact that the actions of the defendant had delayed the claimant's recovery. This approach ties in with the general principle that the function of damages is to compensate the claimant, rather than to punish the defendant.

It can be seen from the discussion on general and special damages and aggravated damages that both these types of damages are compensatory in nature, and that a claimant will not make a profit out of them. Although greater sums may be awarded in the case of aggravated damages, these increased sums only reflect the increased loss or suffering to which the defendant has subjected the claimant. Where truly moderate sums are awarded for aggravated damages, this rationale is unexceptionable but, where much larger sums are awarded, it may be difficult to distinguish between aggravated damages and exemplary damages, as Lord Wilberforce pointed out in *Cassell v Broome* (1972). The distinction is important, because the function of exemplary damages is to punish the defendant, and it is in such situations that one might suggest that the claimant is making a profit out of the damages awarded.

In *Rookes v Barnard* (1964), the House of Lords described those circumstances in which exemplary damages could be recovered in tort. Lord Devlin held that such damages could be awarded only where authorised by statute, for example, s 13(2) of the Reserve and Auxiliary Forces (Protection of Civil Interests) Act 1951, in the case of oppressive, arbitrary or unconstitutional acts by a government servant, or where the defendant has calculated that he will make a profit out of the tort, even if normal compensatory damages are awarded.

These categories have been strictly adhered to. Thus, in *Cassell*, Lord Reid stated that the oppressive, arbitrary or unconstitutional category did not extend to oppressive action by a company. However, in *Holden v Chief Constable of Lancashire* (1987), it was held that exemplary damages could be awarded for unlawful arrest even if there was no oppressive behaviour by the arresting officer, since the category contemplated that the action be oppressive, arbitrary *or* unconstitutional and not oppressive, arbitrary *and* unconstitutional. The last category is illustrated by the facts of *Cassell*, where the defendants published a book containing defamatory statements about the plaintiff. The defendants were aware that the plaintiff intended to sue if the book was published with these statements, but they calculated that this was a risk worth running, as they estimated that the profits that they would make on the sales of the book would outweigh such ordinary compensatory damages. It was held that in such circumstances an award of exemplary damages was appropriate.

In *Cassell*, it was held that exemplary damages are only available in those categories described in *Rookes*, and this whole area has recently been considered in *Kuddus v Chief Constable of Leicestershire Constabulary* (2001). In *Kuddus,* the House of Lords held that the award of exemplary damages was not limited to cases where the cause of action had been recognised before 1964 as justifying such damages, rather the question was whether the facts fell within the categories described by Lord Devlin in *Rookes*. Thus, in *Kuddus,* the House held that exemplary damages could be awarded for the tort of misfeasance in public office.

Thus, it can be seen that the situations in which a claimant can profit from exemplary damages are subject to some limitations. Indeed, the fear that claimants may profit from exemplary damages was stated in *Thompson v Commissioner of Police of the Metropolis* (1997). Here, the Court of Appeal held that limits should be placed on exemplary damages awarded for unlawful and violent conduct by the police, and set an 'absolute maximum' figure.

Another way in which a claimant may profit from an award of damages is where he sues in respect of a consequence of the defendant's conduct for which he is already insured. In such situations, the rule is that insurance benefits are ignored for the purpose of assessing damages (*Bradburn v Great Western Railway* (1874)), and a similar rule applies to charitable donations (*Parry v Cleaver* (1970)). Thus, a person whose house or car is destroyed by a runaway lorry may well make a profit on the damages received if the insurance company forgoes its right of subrogation. Applying the rule in *Bradburn*, the House of Lords held in *Hussain v New Taplow Paper Mills* (1988) that where an employer funded his sick pay scheme through an insurance company, payments so received by an employee should be taken into account in assessing the damages payable in respect of loss of earnings. In contrast, the Court of Appeal in *McCamley v Cammell Laird Shipbuilders* (1990) held that insurance benefits received by an employee did not fall to be taken into account, as the payments in that particular case were in the nature of true insurance benefits. The difference between *Hussain* and *McCamley* depends on whether the court decides, on the facts of the case, that the payments in question are truly sick pay, when they will be deducted, or that they are true insurance benefits, when they will be ignored. However, in *Gaca v Pirelli General plc* (2004), the Court of Appeal overruled *McCamley* and held that insurance benefits received by an employee after an accident at work were no longer to be disregarded unless the employee had contributed to the insurance premium. Thus, if contributions have been made by the employee, the employee may profit from the insurance payment. Any *ex gratia* payments made by the tortfeasor would, however, normally be deducted from the damages awarded. A certain amount of double recovery is allowed in respect of State benefits. By the Social Security (Recovery of Benefits) Act 1997, some State benefits may be deducted from awards of damages, while others are ignored for the purposes of deduction. Thus, no deduction is made against awards for pain and suffering, and only certain specified benefits may be deducted from awards for loss of earnings, cost of care and loss of mobility. Hence, to the extent that the 1997 Act specifies no deduction of State benefits, a double recovery situation exists allowing the claimant to make a profit. In addition, deductions cease after a five year period (s 3 of the 1997 Act).

Another area where double recovery is allowed by statute is in the award of damages under the Fatal Accidents Act 1976. By s 4 of the 1976 Act, any benefits that accrue to the dependants as a result of the death of the deceased are to be disregarded. Thus, in *Pidduck v Eastern Scottish Omnibus* (1990), a widow's pension that was paid to a widow following the death of her husband was held to be non-deductible. The 1976 Act also provides, in s 3(3), that in assessing damages for a fatal accident, the chances of the widow remarrying are to be disregarded. In *Martin v Owen* (1992), it was held that the chance that the parties might have divorced should be taken into account. This conclusion seems rather surprising at first but, when one realises that s 4 of the 1976 Act expressly contemplates double recovery, it will be interpreted strictly to restrict any such double recovery to that stated in the Act. So, a widow who remarried after being awarded damages for loss of dependency could make a profit out of those damages, as could a claimant whose medical condition dramatically improved after an award of damages, either through an unforeseeable medical improvement, or because of an advance in medical science made after the award, as the original award will dispose of the case (*Fitter v Veal* (1701)).

It could also be argued that as, by s 1(1) of the Administration of Justice Act 1982, a sum of £7,500 is paid for loss of a spouse or child, if this sum is paid following the death of a small child, this represents a profit.

Finally, one might consider the position of successful claimants in defamation actions. Where the defendant is a newspaper, juries do seem to forget the principle that the object of awarding damages in tort is to compensate the claimant and not to punish the defendant, and the very large damages that are sometimes awarded against newspapers especially do seem to contain an element of punishment. While the man in the street may well find this quite acceptable, it does represent legally incorrect principles, and claimants who are awarded sums for damages which are in the six and seven figure range are surely making a profit out of their damages.[1]

Think point

1 In this respect, it should be noted that under s 8 of the Courts and Legal Services Act 1990, the court has the power to order a new trial or to substitute another award in any case where the damages awarded by a jury are 'excessive'. In *Rantzen v Mirror Group Newspapers* (1993), the Court of Appeal reduced an award by some 50% and stated that juries could be referred to these substituted awards as establishing norms. In *John v Mirror Group Newspapers Ltd* (1996), it was realised that the *Rantzen* procedure would take time, and the Court of Appeal was prepared to allow juries to be told of awards in personal injury cases. The idea of this is not to promote equality of damages, but to enable a jury to compare a serious libel with, for example, a serious head or spine injury. In time, this may have the effect of reducing damages awarded by juries in defamation cases.

Notes

15 General Defences

Introduction

The general defences to tort are invariably tested by examiners. This may take the form of a specific question on, say, *volenti*, or may form part of another question. Thus, contributory negligence is regularly tested in questions involving a variety of aspects of the tort of negligence. Where contributory negligence is tested, apart from the seat belt guidelines in *Froom v Butcher* (1975), candidates would not be required to estimate figures for any reduction in damages.

Checklist

Candidates must be aware of the following defences:

(a) necessity;

(b) statutory authority;

(c) *volenti*;

(d) illegality; and

(e) contributory negligence.

Question 49

Norman and Mark went out for a social evening using Mark's car. They called at a public house where they both consumed a large amount of drink. Mark then drove Norman home and, due to his intoxicated state, crashed the car against a lamp post. Norman, who was not wearing a seat belt, was thrown through the car windscreen and was severely injured. Rita, who witnessed the accident, went to help Norman and cut her hands badly in so doing.

Advise Norman and Rita of any rights they might have against Mark. Would your advice differ if, rather than going out together, Norman had met Mark in the public house when Norman had had little to drink but Mark was already intoxicated, and Norman had then accepted a lift from Mark?

Answer plan

The following points need to be discussed:

- liability of Mark to Norman;
- defences available to Mark:
  - *volenti* – consideration of case law and statute law;
  - *ex turpi causa*;
  - contributory negligence in accepting the lift;
  - contributory negligence in not wearing a seat belt;
- liability of Mark to Rita;
- defences available to Mark:
  - *volenti*;
  - contributory negligence; and
- effects of Mark's existing intoxication on any defences available to him.

Answer

We must first decide whether Norman could sue Mark and, if so, whether Mark has any defences available to him.

Norman must first show that Mark owes him a duty of care. In those situations where a duty of care has previously been found to exist, there is no need to apply the modern formulation of the test for the existence of a duty of care preferred by the House of Lords in *Caparo Industries plc v Dickman* (1990) or *Murphy v Brentwood District Council* (1990). We could note here the statement of Potts J at first instance in *B v Islington Health Authority* (1991), where he stated that in personal injury cases the duty of care remains as it was pre-*Caparo*, namely, the foresight of a reasonable man (as in *Donoghue v Stevenson* (1932)), a finding that does not appear to have been disturbed on appeal (1992). In fact, a duty of care has been found to exist in a number of cases involving drivers and their passengers, for example, *Nettleship v Weston* (1971) but, even without knowledge of such cases, we could deduce the existence of a duty of care as it is reasonably foreseeable that by driving carelessly, a passenger may suffer injury.

Next, Norman must show that Mark was in breach of this duty, that is, that a reasonable person, or rather a reasonably competent driver, in Mark's position would not have acted in this way (*Blyth v Birmingham Waterworks* (1856); *Nettleship v Weston* (1971)). It seems clear that a reasonable driver would not run into a lamp post, and so Mark is in breach of his duty. Norman will also have to show that this breach caused his injuries, and the 'but for' test in *Cork v Kirby MacLean* (1952) proves the required causal connection. Finally, Norman will have to prove that the damage that he has suffered was not too remote, that is, it must be reasonably foreseeable (*The Wagon Mound (No 1)* (1961)). This should give rise to no problem, as all that Norman will have to show is that some personal injury was foreseeable. He will not have to show that the extent was foreseeable, nor the exact manner in which the injury was caused (*Smith v Leech Brain* (1962); *Hughes v Lord Advocate* (1963)).

Thus, having decided that Mark has been negligent in his conduct towards Norman, we next need to see if any defences are available to Mark. The first possible defence is that of *volenti* on Norman's part, that is, that Norman voluntarily submitted or consented to the risk of injury. To establish that Norman was *volens*, Mark will have to show that Norman was able to choose freely whether to run the risk or not and that there were no constraints acting on his freedom of choice, such as fear of loss of his employment (*Bowater v Rowley Regis Corp* (1944)). In the instant case, no such restraints were acting on Norman. The next point that we must consider is whether there was any agreement between Mark and Norman, whereby Norman agreed to accept the risk of injury. If there was an express agreement that Mark would not be liable to Norman in respect of his negligence then, subject to the provisions of the Unfair Contract Terms Act 1977, that agreement would prevail. There is no evidence on the facts that we are given to suggest such an agreement, so we need to consider whether there was an implied agreement. In cases involving persons who have accepted lifts from persons whom they know to be intoxicated, the courts are usually unwilling to find that the person accepting the lift has impliedly agreed to waive his right to sue the intoxicated driver (*Dann v Hamilton* (1939)). Although an implied agreement was found in *ICI v Shatwell* (1965), that case involved such an obviously dangerous act that it was not difficult for the court to imply an agreement that the two defendants had accepted the risk of any injury following from that most dangerous practice. Given then that there is no agreement, express or implied, between the parties, we next have to consider whether the *volenti* defence could be valid in those circumstances where there is no agreement between the parties. In *Nettleship*, Lord Denning stated that nothing short of an express or implied agreement would suffice to found a defence of *volenti*. However, this view has not been universally accepted. In *Dann*, the court held that *volenti* could apply to those situations where the claimant comes to a situation where a danger has been created by the defendant's negligence (though, on the facts of *Dann*, it was held that *volenti* had not been made out). Also in *Pitts v Hunt* (1990) and *Morris v Murray* (1990), it was held that the defence could apply in appropriate circumstances to passengers who accepted lifts from drivers who were obviously highly intoxicated. Thus, it would seem that, despite *Nettleship*, Mark could raise the defence of *volenti* if he was obviously extremely drunk but, unfortunately for Mark, s 149(3) of the Road Traffic Act 1988 rules out *volenti* in road traffic situations – see *Pitts* and compare *Morris*, which was not a road traffic situation.

Thus, Mark cannot rely on the *volenti* defence, but he may attempt to raise the defence of *ex turpi causa non oritur actio* (which has been expressly held to apply to actions in tort (*Clunis v Camden and Islington Health Authority* (1998))), in that both he and Norman were jointly participating in an illegal activity, namely, driving a motor vehicle whilst under the influence of excess alcohol, contrary to s 4 of the Road Traffic Act 1988. This defence was upheld in *National Coal Board v England* (1954) and *Ashton v Turner* (1981), but there must be a causal connection between the crime and the damage which the claimant has suffered (*National Coal Board*). In *Euro-Diam v Bathurst* (1988), Kerr LJ stated that the defence would apply where it would be an 'affront to the public conscience' to allow the claimant to succeed. This test was also used by Beldam LJ in *Pitts*, but Dillon and Balcombe LJJ preferred to determine whether the claimant's damage was incidental to the unlawful conduct. It is submitted that the defence would fail because there is not the required causal connection between the damage and the crime, as in

National Coal Board and *Ashton*. In *Tinsley v Milligan* (1993), the House of Lords rejected the 'affront to public conscience' test, preferring the test as to whether the claim is based directly on the illegal conduct, and this test has been used by the Court of Appeal in *Cross v Kirkby* (2000) and *Vellino v Chief Constable of Greater Manchester* (2002). Here it does seem that Norman's damage is incidental to Mark's illegal activity. Certainly, in a number of cases, passengers have been allowed to recover in similar situations, for example, *Dann*. In support of this conclusion, one might note the Scottish case of *Weir v Wyper* (1992), where it was held that the *ex turpi causa* defence could not be raised against a passenger who accepted a lift from a driver who she knew possessed only a provisional driving licence. The *ex turpi causa* defence was also given a restricted application in *Revill v Newberry* (1996).

Although the above two defences, which would provide a complete defence to Mark, are not applicable, Mark may be able to raise the defence of contributory negligence to reduce the damages which he will have to pay Norman (s 1(1) of the Law Reform (Contributory Negligence) Act 1945). By s 1(1), where a person suffers damage as the result partly of his own fault and partly of the fault of any other person, his damages will be reduced by such an extent as the court thinks just and equitable, having regard to the first person's share in the responsibility for the damage. To raise this defence, Mark will have to show that Norman was careless for his own safety (*Davies v Swan Motor Co* (1949)). In *Jones v Livox Quarries Ltd* (1952), Lord Denning said that 'a person is guilty of contributory negligence if he ought reasonably to have foreseen that, if he did not act as a reasonable prudent man, he might hurt himself; and in his reckonings he must take into account the possibility of others being careless'. On this basis, Norman has been careless in accepting a lift from a driver whom he knows to be intoxicated (*Dann*; *Pitts* at first instance; *Owens v Brimmel* (1977)). Thus, any damages Norman receives will be reduced due to this particular act of contributory negligence. In addition, we are told that Norman was not wearing a seat belt at the time of the crash and, as we are told that he was thrown through the windscreen, it is clear that if he had been wearing a seat belt, the extent of his injuries would have been reduced. Although Norman's act in not wearing a seat belt did not contribute to the accident, it has contributed to the extent of the damage he has suffered, and so his damages will be further reduced (*Froom v Butcher* (1975)).

Hence, Norman should be advised that he can recover damages from Mark, but these damages will be reduced to take into account his contributory negligence.

Turning now to Rita, she is a rescuer and can sue Mark (*Haynes v Harwood* (1935)). Mark will almost certainly be unsuccessful in attempting to raise the defence of *volenti* against Rita (*Haynes v Harwood* (1935); *Chadwick v British Transport Commission* (1967)). The only situation in which a rescuer will be held to be *volens* is where a rescue is attempted in circumstances in which there is no real danger (*Cutler v United Dairies* (1933)), which is not the case here. Mark may try to run the defence of contributory negligence against Rita in an attempt to reduce any damages payable to her, but the courts are reluctant to find rescuers guilty of contributory negligence. This has been done where the circumstances warrant it (*Harrison v British Railways Board* (1981)) but, in judging whether or not the rescuer has been careless for her own safety, the courts take into account the fact that by the negligence of the defendant, the claimant may have been placed in an emergency, and will be sympathetic to a claimant who makes a wrong decision in the agony of the moment (*Jones v Boyce* (1816)). Thus, on the facts that we are given, it seems unlikely that a finding of contributory negligence would be made against Rita, who could recover in full against Mark.

If Norman had met Mark in the public house when Norman had had little to drink, but Mark was already intoxicated, then *prima facie* it would be easier for the court to find that Norman was *volens* to the risks of being a passenger in Mark's car. In *Dann*, it was said that if the drunkenness of the driver was extreme, the *volenti* defence might apply. However, even if this were to apply to Mark, s 149(3) would still render the defence invalid.

Notes

Question 50

'In practice, the so called "general defences" in tort are so confused that a defendant would be ill advised to rely on them.'

Discuss.

Answer plan

This is a straightforward essay on the general defences in tort, but candidates should not make the error of merely listing and describing the general defences – a discussion of the uncertainties and undecided points appertaining to each defence is what the examiner is looking for in this question.

Bearing this in mind, the following points need to be discussed:

- necessity;
- statutory authority;
- consent;
- illegality; and
- limitation.

Answer

The defence of necessity exists, as, for example, in the cases of *Cope v Sharpe* (1912) and *R v Bournewood Community and Mental Health NHS Trust* (1998), but it is not favoured by the courts. Thus, the defence was not allowed by the Court of Appeal in *Southwark Borough Council v Williams* (1971), where Lord Denning said that to allow it 'would be an excuse for all sorts of wrongdoing', and a similar approach was taken in the House of Lords in *Burmah Oil v Lord Advocate* (1965).

It seems, therefore, that the statement under discussion is accurate as regards this general defence, insofar as the extent of the defence has yet to be defined by the courts.

The defence of statutory authority, although perhaps not a common defence, certainly exists. The rule is that where a statute authorises an act which would otherwise be actionable, no action will lie in respect of that act. Additionally, no action will lie as regards any necessary consequences of that act. By 'necessary consequences', we mean those consequences that cannot be avoided by the exercise of proper care and skill; if any consequences are so avoidable, then an action will lie in respect of them. In other words, statutory authority is authority to carry out the relevant activities carefully and properly, not authority to carry out the activities carelessly, for that could not have been the intention of Parliament.

The courts nowadays construe statutes liberally rather than strictly, so the House of Lords in *Allen v Gulf Oil Refining* (1981) held that where a statute gave the defendant power to build a refinery, it had by implication authorised its operation, and so no action would lie in respect of any necessary consequences attaching only to its operation. Avoidable consequences may arise not only from careless operation of a facility but also from a negligent choice of mode of carrying out the authorised act. This can be seen in the decision of the House of Lords in *Tate and Lyle v Greater London Council* (1983), where their Lordships held that the defendant could be sued in respect of their negligent choice, rather than operation, of a mode of carrying out their statutory powers.

It can thus be seen that in the appropriate circumstances, the defence of statutory authority is a reliable defence, in that its scope has been considered and defined by the courts, and it is a defence upon which a defendant could be advised to rely. Thus, the statement under discussion is not an accurate reflection of the law as regards this defence.

The defence of consent, or of *volenti non fit injuria*, is well-established in terms of legal theory, but we need to consider how effective it is as a practical defence in terms of its extent and boundaries being defined. As regards the infliction of intentional harm, consent is a valid defence that causes few problems. A moment's reflection will show how it could be a valid defence to an action for trespass to land or trespass to the person. However, when one considers the infliction of accidental harm, a number of problems can arise in practice in establishing this defence.

It must, for example, be shown that the claimant voluntarily submitted to the risk of injury (*Bowater v Rowley Regis Corp* (1944)), and that there were no constraints operating on the claimant, such as the fear of losing his job. If no such constraints operate and an employee chooses to adopt a dangerous method of working which causes him damage, the courts may find that he was *volens* to the injury (*ICI v Shatwell* (1965)), although in practice, especially with employees, the courts are reluctant to make a finding of *volenti* and prefer instead to make a finding of contributory negligence. In particular, a finding of *volenti* is rarely made against rescuers (*Chadwick v British Transport Commission* (1967)) or persons who are of unsound mind (*Kirkham v Chief Constable of Greater Manchester* (1990)), although intoxication does not necessarily rule out a finding of *volenti* (*Morris v Murray* (1990)).

However, there is a certain amount of confusion in the authorities as to whether agreement is an essential ingredient of *volenti*. In *Nettleship v Weston* (1971), Lord Denning stated that either express or implied agreement was an essential ingredient of *volenti* but, in *Dann v Hamilton* (1939), it was held not to be essential. *Dann* has been subject to a certain amount of criticism (see, for example, *Pitts v Hunt* (1990) at first instance), but the majority view seems to be that agreement is not necessary.

Another problem that may arise in raising this defence is that if there was an express agreement between the parties that the claimant will run the risk of any injury, that agreement may be caught by the Unfair Contract Terms Act 1977. In particular, s 2(1) provides that a person cannot exclude or restrict his liability for death or personal injury resulting from negligence and, by s 2(2), any restriction for other loss or damage must satisfy the requirement of reasonableness.

Another very important restriction on the *volenti* defence is that imposed by s 149 of the Road Traffic Act 1988, which renders void any agreement or notice purporting to exclude liability in situations where insurance is compulsory. Thus, *volenti* will never be a valid defence in road traffic situations (*Pitts*), but it can be raised in cases of non-road transport, for example, aircraft, as in *Morris v Murray* (1990).

It can therefore be seen that the *volenti* defence does have areas that are capable of giving rise to confusion. It is still not certain whether agreement is essential, although in appropriate cases the courts would probably be ready to find implied agreement. Where s 2(2) of the Unfair Contract Terms Act 1977 applies, there is of course the additional problem of deciding whether the particular exclusion clause is reasonable or not. Thus, it is submitted that in relying on this defence, there may be confusion as to the circumstances in which it applies and, where it involves the existence of an exclusion clause, its scope may not be easy to determine *a priori*.

Illegality is a general defence, and is sometimes described as the *ex turpi causa non oritur actio* defence. It has been expressly held to apply to actions in tort (*Clunis v Camden and Islington Health Authority* (1998)). Its most obvious application is where the parties have been participating in a joint criminal activity (*National Coal Board v England*

(1954); *Ashton v Turner* (1981)). It is, however, not confined to criminal activities, but is of much wider scope (*Euro-Diam Ltd v Bathurst* (1988)). The precise criterion which activates the defence seems to be a little uncertain – in *Euro-Diam*, the affront to public conscience test was used, that is, that it would be an affront to the public conscience to grant the claimant relief, as it would appear to assist his illegal conduct (see also *Thackwell v Barclays Bank* (1986); *Saunders v Edwards* (1987)). This test was also adopted by Beldam LJ in *Pitts*, while Dillon and Balcombe LJJ preferred to base their decisions on whether the claimant's claim was based directly on his illegal conduct or whether the illegal conduct was merely incidental. More recently, however, the House of Lords in *Tinsley v Milligan* (1993) rejected the affront to public conscience test, and the test would now seem to be whether the claim is based directly on the illegal conduct. This was the test used by the Court of Appeal in *Cross v Kirkby* (2000) and *Vellino v Chief Constable of Greater Manchester* (2002). In *Tinsley*, the plaintiff and defendant bought a house in which they lived together, but which was held in the sole name of the plaintiff. The purpose of this arrangement was to perpetrate a fraud on the Department of Social Security, and such a fraud was in fact carried out by both parties. Following a disagreement, the plaintiff brought proceedings claiming to be the sole owner of the property and the defendant counter-claimed that she was entitled to a one-half share in the property. The plaintiff failed in her claim, the counter-claim being allowed, both at first instance and in the Court of Appeal. Her appeal was based on the ground that the defendant could not succeed, because of the *ex turpi* doctrine. The House of Lords dismissed the appeal on the grounds that the defendant was entitled to recover her share in the property, as she was not forced to rely on the illegality, even though the title on which she relied was acquired through an illegal transaction. Although the House rejected the affront to public conscience test and instead decided whether the claim was based directly on the illegal conduct, which might at first glance appear to be a simpler test and easier to apply and less confusing, it is worth noting that, based on this test, the House of Lords only managed to reach a bare majority verdict of three to two in this case. It thus seems that although the direct action test may appear to be clearer than the old affront to public conscience test, it is by no means easy to predict the outcome in all cases, and so it could be said that some confusion still exists in relying on this defence.

A final general defence is that afforded by the Limitation Act 1980. Section 2 provides that an action in tort cannot be brought more than six years after the cause of action accrued, or three years in the case of personal injury. Although these time periods will often provide a certain defence, they are subject to some exceptions and, in particular, for personal injury actions, the court has under s 33 a wide discretion to allow such an action to proceed out of time. However, this discretion can only be exercised in personal injury cases arising from negligence; where the personal injury results from a deliberate assault, then the six year period provided by s 11 of the Limitation Act 1980 applies and cannot be extended (*Stubbings v Webb* (1993)).

Thus, apart from the matter of the court's discretion under s 33, the defence of limitation is defined with certainty, and there are of course many authorities on the application of s 33. Overall, therefore, it cannot be said that confusion exists with respect to the application of this defence.

Returning now to the original statement, it can be seen that each of the general defences is subject to an element of confusion or uncertainty. However, it is submitted that to state that they are so confused that a defendant would be ill advised to rely on them is an exaggeration that cannot be borne out by a study of the relevant authorities, and that there are a number of situations in which a defendant would be well-advised to rely on a general defence.

Notes

Index

Ambulance services, liability of **37–38, 42**
Animals **137–46**
- Animals Act 1971 . . . . 137–41, 143, 168–69
- causation 142, 144–45
- characteristics of animals 138, 141, 144, 169
- common law 137, 139
- contributory negligence 141, 144
- damage, liability for 137–45, 169
- dangerous and non-dangerous species 137–38, 141, 143–44
- dogs 137–42, 166–67
- foreseeability 139, 142–44
- guard dogs 139, 144
- keepers 137–39, 141, 144, 169
- negligence 142, 168
- *novus actus interveniens* 139, 142, 144–45
- nuisance 145
- personal injuries 137–39, 145
- reasonableness 139, 142
- *Rylands v Fletcher* 139, 142, 145
- straying livestock 137
- third parties 142, 144–45
- trespassers 139, 144
- *volenti non fit injuria* 139, 141, 144

Appropriate adults, psychiatric harm and **41–42**
Arrest **164, 165–66, 172–74**
Assault **164, 171, 172–75**

Bank of England, misfeasance in public office and **26**
Barking dogs **101–05**
Battery **165, 171, 173–75**
Bereavement damages **191–92**
***Bolam* test** **48**
Breach of contract, inducement of **179, 180, 181–82**
Breach of duty **43–44**
- *See also* Breach of statutory duty
- employers' liability 46–49
- medical negligence 48–49
- reasonableness 44
- *res ipsa loquitur* 45
- road accidents 44–45
- *Rylands v Fletcher* 122
- standard of care 44

Breach of statutory duty **53–61**
- *See also* Local authorities, breach of statutory duty and
- but for test 56
- clothing, maintenance of protective 53–56
- delegation 56, 72–73
- employers' liability 53–56, 70, 72–73
- enforcement 53
- equipment, maintenance of protective 55–56
- fencing machinery 54
- fines 54–55
- foreseeability 56
- harm 53–55
- *novus actus interveniens* 56
- nuisance 133
- reasonableness 54, 56
- safe system of work 55–56
- standard of care 54
- vicarious liability 53–54, 56
- *X v Bedfordshire CC* 57–58

But for test **43, 49–52**
- breach of statutory duty 45, 56
- employers' liability 47–48
- medical negligence 50
- occupiers' liability 89, 90
- pre-existing conditions 50
- product liability 84
- remoteness 44–45
- road accidents 44–45
- vicarious liability 2, 6

Carelessness **1, 5, 14, 38, 65, 68**
Causation
- *See also* But for test
- animals 142, 144–45
- egg shell skull rule 48
- employers' liability 47–48
- material contribution test 52
- medical negligence 48
- misfeasance in public office 26
- *novus actus interveniens* 45, 51–52
- product liability 77–78, 80, 84–85
- reasonableness 51–52
- *Rylands v Fletcher* 131

third parties . . . 51–52
trespass . . . 164–65
vicarious liability . . . 64–65

Charitable donations, damages and . . . 195

Children
local authorities . . . 58–61
occupiers' liability . . . 87–89, 96
sexual abuse . . . 31, 34–35, 58–61
special educational needs, children with . . . 58–61

Clothing, maintenance of protective . . . 53–56

Coastguards, liability of . . . 37–38, 42

Competition . . . 178

Consent
See Volenti non fit injuria

Conspiracy to commit unlawful acts . . . 178–79, 181, 184–86

Contractual relations, interference with . . . 186

Contributory negligence
animals . . . 141, 144
damages, reduction in . . . 4, 200
occupiers' liability . . . 90–91
rescuers . . . 200
road accidents . . . 200
seat belts . . . 1, 4
vicarious liability . . . 2, 4

Conversion . . . 168, 170–71

Damages . . . 189–96
aggravated . . . 193–94
assessment of . . . 189–93
bereavement . . . 191–92
charitable donations . . . 195
contributory negligence . . . 4, 200
death . . . 189, 191–92, 195
defamation . . . 150, 156, 161–62, 196
dependants . . . 191–92, 195
exemplary . . . 194
fire . . . 134–35
future loss . . . 190–92
insurance . . . 195
interest . . . 191
life expectancy . . . 190–92
loss of earnings . . . 190–93, 195
lump sums . . . 190
multipliers . . . 190, 192
nuisance . . . 103, 104, 111, 129
oppressive, arbitrary and unconstitutional acts . . . 194
pain and suffering . . . 191, 195
pension rights, loss of . . . 190–91
personal injuries . . . 189
profits . . . 192–96
provisional awards . . . 190
road accidents . . . 189–92
Rylands v Fletcher . . . 134–35
social security benefits, deduction from . . . 191, 195
structured settlements . . . 190
trespass . . . 164, 168, 171

Death, damages and . . . 189, 191–92, 195

Deceit . . . 181–83, 186–87

Defamation . . . 147–62
amends, offers of . . . 161
apologies . . . 150, 156, 161
class defamation . . . 148–49, 155–56
commentary . . . 155–56
consent . . . 159
corrections . . . 156, 161
damages . . . 150, 156, 161–62, 196
Defamation Act 1996 . . . 147
economic torts . . . 179–80, 183
fair comment . . . 148–53, 156–57
fast track procedure . . . 150
foreseeability . . . 157, 159
freedom of expression . . . 158–61
innuendo . . . 148, 152, 155
juries, damages and . . . 196
justification . . . 148–49, 152–53, 156–57, 159
libel . . . 148, 152
malice . . . 149–50, 152–53, 157, 159, 162, 175
negligence . . . 161
negligent misstatements . . . 152
opinions . . . 149, 153, 157, 161
political expression . . . 158, 161
public interest . . . 149, 152–53, 157
publication . . . 149, 152, 155, 157, 159, 161, 168, 174–75
qualified privilege . . . 151–54, 159–61, 165, 168, 173, 175
relationship cases . . . 153, 175
repetition . . . 155, 157, 159
slander . . . 155–56, 174, 180
special damage . . . 148, 150, 151, 155, 174, 180, 183
standing . . . 155–56
time limits . . . 150
truth . . . 156–57

Defective products
See Product liability

Delegation
breach of statutory duty . . . 56, 72–73
employers' liability . . . 64, 71
vicarious liability . . . 9–10

Dependants, damages and 191–92, 195
Dogs 101–05, 137–42, 166–67
Drunk driving 197–99
Duty of care 13–42, 198
See also Breach of duty
Anns test 13–16
Caparo case 47, 56, 59
carelessness 14
economic loss 14
emergency services, liability of 37–42
employers' liability 67
foreseeability 15, 44, 47, 56
incremental approach 13–15
negligent misstatements 17–23
neighbour test 14
nervous shock 14
novus actus interveniens 19
occupiers' liability 88–89, 94–99
passengers 2–8
product liability 83
proximity 14–16
reasonableness 15, 19, 44, 47, 56
road accidents 198
two stage test 15–16
vicarious liability 1–9, 64

Economic loss
Junior Books case 25, 78, 82
misfeasance in public office 25–27
negligence 25
nuisance 108–09, 114, 118
product liability 78, 82
Rylands v Fletcher 123, 124
Economic torts 177–87
competition 178
conspiracy to commit unlawful acts 178–79, 181, 184–85
contractual relations, interference with 186
deceit 181–83, 186–87
defamation 179–80, 183
duty of care 14
inducement of breach of contract 179, 180, 181–82, 185–86
intent 184–87
intimidation 180, 186
malicious falsehood 178–80, 183, 187
mere puffs or opinions 183
trade, interference with 179, 180, 182, 186–87
Egg shell skull rule 5, 7, 48, 65, 84
Emergency services, liability of 37–42
See also Police
ambulance services 37–38, 42
carelessness 38
coastguards 37–38, 42
duty of care 37–42
fair trials, immunity and 40
fire services 38–39, 41–42
omissions 41
proximity 41
public policy 41
Emissions 105–08, 121–24, 131–33
Employers' liability 46–49, 63–73
breach of duty 46–49
breach of statutory duty 53–56, 70, 72–73
but for test 47–48
causation 47–48
delegation 64, 71
duty of care 67
equipment, defective 46–47, 66, 71, 72
fellow employees, to 71–72
foreseeability 48, 67–68, 69
independent contractors 68, 69
personal duty 71–73
plant and machinery 71
reasonableness 67, 71–72
remoteness 48
safe place of work 64, 68, 69, 71, 72
safe system of work 71
secondary liability 64, 69
volenti non fit injuria 203
Equipment 46–47, 55–56, 66, 71, 72
Escapes from police custody, injuries during 38
***Ex turpi causa* test 19, 38, 90–91, 94, 199–200, 203–04**
Exclusion of liability 95–99, 199, 203

Fair trials 40, 60–61
False imprisonment 164, 165–68, 173–75
Family life, right to respect for 61, 104, 108–09, 113, 118–19, 123–24
Fencing machinery 54
Finders 169–71
Fire
damages 134–35
fire services, liability of 38–39, 41–42
negligence 132, 133–35
nuisance 132
Rylands v Fletcher 132, 134–35
Foreseeability
animals 139, 142–44
breach of statutory duty 56
defamation 157, 159
duty of care 15, 44, 47, 56
employers' liability 48, 67–68, 69

misfeasance in public office . . . 26
negligent misstatements . . . 19, 20–22
nervous shock . . . 28–29, 32–35
nuisance . . . 111–12, 115, 117, 133
occupiers' liability . . . 90, 96
personal injuries . . . 198–99
product liability . . . 78–79, 80–85
remoteness . . . 45
road accidents . . . 198–200
Rylands v Fletcher . . . 107, 113, 122, 126–28, 131
vicarious liability . . . 2, 5, 7, 9, 64–66

Freedom of expression, defamation and . . . 158–61

Goods
See Product liability; Trespass to goods

Government departments, misfeasance in public office and . . . 25

Guard dogs . . . 139, 144

Human Rights Act 1998
local authorities, breach of statutory duty and . . . 57, 60–61
nuisance . . . 101, 104–06, 108–10, 113, 118–19
Rylands v Fletcher . . . 122, 123–24

Illegality . . . 25–26, 90, 94, 203–04

Immunity . . . 38, 40–42, 60–61

Independent contractors
employers' liability . . . 68, 69
occupiers' liability . . . 68, 87–90, 92, 96–97
Rylands v Fletcher . . . 127–28
vicarious liability . . . 8–11

Inducement of breach of contract . . . 179, 180, 181–82

Informers, protection of . . . 40

Injunctions, nuisance and . . . 103, 107, 111, 116

Intervening acts
See Novus actus interveniens

Intimidation . . . 180, 186

Licences, occupiers' liability and . . . 89

Life, right to . . . 108, 118, 123

Livestock, straying . . . 137

Local authorities
See also Local authorities, breach of statutory duty and
misfeasance in public office . . . 25
nuisance . . . 104, 108–13, 118
Rylands v Fletcher . . . 124

Local authorities, breach of statutory duty and . . . 57–61
discretion . . . 59–60
fair trials . . . 60–61
family life, right to respect for . . . 61
Human Rights Act 1998 . . . 57, 60–61
immunity . . . 60–61
policy and operational decisions . . . 59
sexual abuse . . . 58–61
special educational needs, children with . . . 58–61
ultra vires . . . 59
vicarious liability . . . 60, 61

Maintenance . . . 53–56

Malice
defamation . . . 149–50, 152–53, 157, 159, 162, 175
misfeasance in public office . . . 25–26
nuisance . . . 104, 116
Rylands v Fletcher . . . 127
targeted malice limb . . . 25–26

Malicious falsehood . . . 178–80, 183, 187

Medical negligence
allergy tests . . . 48–49
Bolam test . . . 48
breach of duty . . . 48–49
but for test . . . 50
causation . . . 48, 50
novus actus interveniens . . . 45
pre-existing conditions . . . 50
standard of care . . . 48–49

Mere puffs or opinions . . . 183

Misfeasance in public office . . . 24–27
bad faith . . . 26
Bank of England . . . 26
causation . . . 26
economic loss . . . 25–27
foreseeability . . . 26
government departments . . . 25
illegality limb . . . 25–26
local authorities . . . 25
mental element . . . 25–26
negligence, advantages over . . . 24–25
proximity . . . 26–27
public policy . . . 26–27
special damage . . . 26
targeted malice limb . . . 25–26
Three Rivers DC case . . . 24, 25–26

Necessity . . . 202

Negligence
See also Contributory negligence; Negligent misstatements
animals . . . 142, 168

causes of action 25
collateral 9–10, 68
damage to property. 123
defamation 161
economic loss. 25
fire 132, 133–35
medical. 45, 48–50
misfeasance in public office. 24–25
novus actus interveniens. 45
occupiers' liability 90–91
proximity. 25
res ipsa loquitur 123
Rylands v Fletcher. 129, 131–32, 133–35

Negligent misstatements
Caparo test. 20–22
class, size of. 22
defamation 152
detriment 18, 23
duty of care. 17–23
foreseeability. 19, 20–22
general circulation,
statements in 20
guidelines. 21
Hedley Byrne test 17–18, 20, 22
independent advice. 21–22
knowledge. 21, 22–23
proximity 20–22
purpose of statement 21–23
reasonableness. 17–19, 22–23
references, reliance on 18
reliance 17–18, 20, 22
skill, possession of special 17, 22
social relationships 18, 22
special relationships 17, 22–23
voluntary assumption
of responsibility 18, 23

Nervous shock 27–37
aftermath. 28–31, 33–35
Alcock case. 28, 32–36
bystanders 29, 33, 36
duty of care 14
employee/employer
relationship, effect of 28, 29,
30, 35–36
family relationships 28–29, 32, 34
foreseeability 28–29, 32–35
grief and sorrow 29, 33
legislation 32
love and affection,
close relationships of 28, 30–31,
33–34, 36
McLoughlin decision 32
primary victims 28, 29–30, 33, 35–36
police 35–36
property damage. 33–34
proximity 28–29, 31, 32–36
reasonableness 28–30, 32–34, 36
rescuers 28, 30, 35–36
secondary victims 29–30, 35
sexual abuse. 31, 34–35
television broadcasting 33
W v Essex CC 34–35
White case 28, 32, 35–36

**Noise nuisance. 101–05, 107,
109–13, 116, 118**

**Notices 1, 4, 88–89,
91–94, 96, 98**

Novus actus interveniens
animals 139, 142, 144–45
breach of statutory duty 56
causation 45, 51–52
duty of care 19
medical negligence. 45
product liability 85
reasonableness 19
trespass to the person. 164–65

Nuisance. 101–19
animals. 101–05, 145
but for test 133
damage to property. 133
damages 103, 104, 111, 129
dogs, barking 101–05
duration of interference 103, 115–16, 132
economic loss 108–09, 114, 118
emissions 105–08, 131–33
enjoyment and use of
land, interference with 102–04, 106, 110,
114–15, 117,
132–33, 168
fire 132
foreseeability 111–12, 115,
117, 133
Human Rights Act 1998 . . . . 101, 104–06,
108–10, 113, 118–19
injunctions 103, 107, 111, 116
interests in land. 101–02, 106–09, 111,
113, 115, 118,
123, 133, 145
isolated events 103, 106–07, 127–28,
132–33, 145
landlords. 101–02
life, right to 108, 118
local authorities 104, 108–13, 118
malice. 104, 116
neighbourhood,
character of the 103–07, 110–11, 116, 133
noise 101–05, 107,
109–13, 116, 118
peaceful enjoyment
of possessions 108, 118
personal injuries 109, 114, 117–18
private 101–02, 107,
114–19, 132
private and family life,
right to respect for. 104, 108–09,
113, 118–19

proximity . . . 108
public . . . 106–07, 111, 123
reasonableness . . . 102–04, 106, 108, 110–12, 114–17, 119, 133
Rylands v Fletcher . . . 125–29, 131–33
sensitivity . . . 103, 106–07, 111, 116
smoke . . . 131–33
statutory . . . 104
television signals, interfering with . . . 102, 104, 115, 133
third parties . . . 129
trespassers . . . 117
utility of conduct . . . 106, 116

Occupiers' liability . . . 87–99
but for test . . . 89, 90
children . . . 87–89, 96
common law . . . 88, 92, 95
contributory negligence . . . 90–91
Donoghue v Folkestone Properties . . . 95
duty of care . . . 88–89, 94–99
ex turpi causa rule . . . 90–91, 94
exclusion of liability . . . 95–99
foreseeability . . . 90, 96
illegality . . . 90, 94
independent contractors . . . 68, 87–90, 92, 96–97
knowledge . . . 97
licences . . . 89
non-visitors, visitors becoming . . . 87–93, 95–99
notices . . . 88–89, 91–94, 96, 98
Occupiers' Liability Act 1957 . . . 87–88, 92, 93–94, 96–97
Occupiers' Liability Act 1984 . . . 87, 90, 94
permission . . . 90
reasonableness . . . 88–90, 92–98
repairs . . . 87–89, 97
Tomlinson v Congleton BC . . . 95
trespassers . . . 93–94, 97–99
unfair contract terms . . . 98
vicarious liability . . . 11
visitors . . . 87–93, 95–99
warnings . . . 88–89, 91–94, 96, 98
Omissions . . . 41
Opinions . . . 149, 153, 157, 161, 183

Passengers . . . 1, 2–8, 199, 200
Peaceful enjoyment of possessions . . . 108, 118, 123
Personal injuries
animals . . . 137–39, 145
damages . . . 189
escapes from custody . . . 38
foreseeability . . . 198–99
nuisance . . . 109, 114, 117–18
Rylands v Fletcher . . . 117–18, 124, 131
time limits . . . 204
Plant and machinery . . . 71
Police . . . 37–40, 42
appropriate adults, duty to . . . 41–42
arrest . . . 164, 165–66, 172–74
escapes from custody, injuries during . . . 38
false imprisonment . . . 164, 165–68, 173–75
immunity . . . 38, 40–42
informers, protection of . . . 40
nervous shock, appropriate adults and . . . 35–36, 40–41
proximity . . . 39–40
public policy . . . 38–40
special relationships . . . 40
suicide in custody . . . 40
Practical jokes . . . 71–72
Private life, right to respect for . . . 104, 108–09, 113, 118–19, 123–24
Product liability . . . 75–86
but for test . . . 84
causation . . . 77–78, 80, 84–85
common law . . . 75–85
Consumer Protection Act 1987 . . . 75–81, 83, 85
damage to property . . . 77, 79–82
duty of care . . . 83
economic loss . . . 78, 82
egg shell skull rule . . . 84
foreseeability . . . 78–79, 80–85
intermediate examination . . . 77, 84
manufacturers . . . 75–86
novus actus interveniens . . . 85
precautions . . . 83–84
Product Liability Directive . . . 76
producers . . . 76, 80, 85
reasonableness . . . 77–78, 80–81, 83–85
remoteness . . . 81
retailers . . . 81
scientific and technical knowledge . . . 76–79, 85
standard of care . . . 83–84
suppliers . . . 76, 80
warnings and instructions . . . 76–77, 84–85
Proximity
duty of care . . . 14–16
emergency services, liability of . . . 41
misfeasance in public office . . . 26–27
negligence . . . 25

negligent misstatements 20–22
nervous shock 28–29, 31, 32–36
nuisance 108
police 39–40

Psychiatric harm
See Nervous shock

Public conscience test 203–04

References, reliance on 18

Remoteness 43
but for test. 44–45
employers' liability 48
foreseeability 45
product liability 81
Rylands v Fletcher 131
vicarious liability 2, 65–66

Res ipsa loquitur **. 2, 45, 123, 131**

Rescuers. 28, 30, 35–36, 200, 203

Road accidents
breach of duty. 44–45
but for test. 44–45
contributory negligence 200
damages. 189–92
drink driving 197–99
duty of care 198
ex turpi causa rule. 199–200
foreseeability. 198–200
passengers, duty to 1, 2–8, 199, 200
seat belts 200
standard of care 44
vicarious liability 2–8
volenti non fit injuria 1, 4, 199, 200, 201, 203

Rylands v Fletcher **. 106, 107, 110, 111, 121–35**
accumulation. 125, 127
animals 139, 142, 145
breach of duty. 122
Cambridge Water case 123, 126
causation 131
damage to property. 122–23, 127
damages. 134–35
dangerous or
mischievous things. 122, 125–29, 134
economic loss 123, 124
emissions 108, 121–24
enjoyment and use of land,
interference with 132
escapes from land 107–08, 111, 121–22, 125–31, 134
fire 132–35
Fires Prevention
(Metropolis) Act 1774 . . . . 132, 135
foreseeability 107, 113, 122, 126–28, 131
Human Rights Act 1998. . . . . 122, 123–24
independent contractors 127–28
interests in land 123, 124, 126, 128, 131, 134
isolated events 127–28
life, right to 123
local authorities. 124
malice. 127
negligence. 129, 131–35
non-natural user 107–08, 111–12, 121–22, 125–27, 129–31, 134
nuisance 125–29, 131–33
ordinary use test 122, 126, 129
peaceful enjoyment
of possessions 123
personal injuries 117–18, 124, 131
private and family life,
right to respect for 123–24
reasonableness 113, 122, 126, 127
remoteness. 131
res ipsa loquitur 131
smoke. 132
statutory authority, defence of 121–23
third parties. 121, 131
Transco v Stockport MBC 121–22
trespassers. 131

Safe place of work. 64, 68, 69, 71, 72

Safe system of work 55–56, 71

Seat belts, failure to wear 1, 4, 197, 200

Self-defence 174

Sexual abuse 31, 34–35, 58–61

Smoke, nuisance and 131–33

**Social security benefits,
deduction from damages of 191, 195**

Special educational needs 58–61

Standard of care. 44, 48–49, 54, 83–84

Statutory authority, defence of 121–23, 202

Statutory duty
See Breach of statutory duty

Straying livestock. 137

Structured settlements. 190

Suicide in custody 40

**Television signals,
interference with 102, 104, 115, 133**

Time limits. 150, 204

Trade, interference with 179, 180, 182, 186–87

Trespass to goods 169–72
conversion. 168, 170–71
damages. 168, 171
finders. 169–71
strict liability 170
vicarious liability 170–71

Trespass to land **166–68**
animals 139, 144
causation 165
dogs 166–67
false imprisonment, police and 167–68
nuisance 117
occupiers' liability 93–94, 97–99
Rylands v Fletcher 131
volenti non fit injuria 203
Trespass to the person **163–66, 170**
arrest 164, 165–66, 172–74
assault 164, 171, 172–75
battery 165, 171, 173–75
causation 164–65
damages 164
false imprisonment,
police and 164, 165–66, 173–75
novus actus interveniens 164–65
permission 164
self-defence 174
touching 173
visitors 164
volenti non fit injuria 203

Ultra vires **59**
Unfair contract terms **4, 98, 199, 203**

Vicarious liability **1–11, 47, 64, 70–71**
authorised acts, wrong and
unauthorised mode of 2–4, 6–7, 9, 65, 71–72
breach of statutory duty 53–54, 56, 60, 61
business integration test 10
but for test 2, 6
careless and deliberate acts 1, 5, 65, 68
causation 64–65
collateral negligence 9–10, 68
contributory negligence 2, 4
control test 9–10
course of employment 1, 2–7, 9, 64–65, 68, 69, 72
definition of employer
and employee 1
delegation 9–10
duty of care 1–9, 64
egg shell skull rule 5, 7, 65
employee status, test for 9–10
foreseeability 2, 5, 7, 9, 64–66
frolics and detours 1, 3–7
independent contractors 8–11
local authorities, breach of
statutory duty and 60, 61
non-delegable duty 9–10, 68, 69
notices 1, 4
occupiers' liability 11
passengers, duty of care to 2–8
practical jokes 71–72
prohibition 2
reasonableness 2, 5–6, 9–11, 64–65
remoteness 2, 65–66
res ipsa loquitur 2
road accidents 2–8
safe place of work 69
seat belts, failure to wear 1, 4
trespass to goods 170–71
volenti non fit injuria 2, 4, 5, 7
Visitors **87–93,**
95–99, 164
Volenti non fit injuria **19, 199, 203**
agreement, role of 203
animals 139, 141, 144
employers' liability 203
notices, awareness of 4
passengers 1, 4, 199, 200
rescuers 200, 203
road accidents 4, 199, 201, 203
trespass to land 203
trespass to the person 203
unfair contract terms 203
vicarious liability 2, 4, 5, 7
**Voluntary assumption
of responsibility** **18, 23**

Warnings and instructions
occupiers' liability 88–89, 91–94, 96, 98
product liability 76–77, 84–85